Resonant Memory

Essays on the Mathematical Laws of History

Resonant Memory

Essays on the Mathematical Laws of History

Michael Gfoeller

Academica Press
Washington

Library of Congress Cataloging-in-Publication Data
Names: Gfoeller, Michael (author)
Title: Resonant memory : essays on the mathematical laws of history | Gfoeller, Michael.
Description: Washington : Academica Press, 2026. | Includes references.
Identifiers: LCCN 2026936402 | ISBN 9781680536454 (hardcover) | 9781680536461 (e-book)

To Kurt v. Klein, adventurer, comrade, and distinguished editor of this and many other volumes of scholarship.

To my wife Tatiana, my muse and inspiration in all things.

Contents

Foreword

Resonant Memory

This book is an inquiry into the mathematical laws that govern history. The confluence of two revolutionary conceptions lies at its heart. On one side stands the theory of morphic resonance, a hypothesis that has grown steadily through decades of research, experimentation, and reflection. On the other side rises a bold extension of spacetime geometry: the proposal that time itself is irreducibly three-dimensional, and that the true arena of physical reality is therefore a six-dimensional manifold in which three dimensions of space are partnered by three dimensions of time. The first idea emerges from biology and reaches toward the foundations of nature's habits; the second arises within theoretical physics and reaches toward a final unification of gravity with the quantum domain. At first glance they appear unrelated, yet when placed in proximity they illuminate one another with startling clarity.

The theory of morphic resonance begins with a simple observation: nature is habitual. Forms recur, behaviours repeat, crystals grow according to patterns that have appeared countless times before. The laws of physics, as usually understood, cannot account for this cumulative character of natural form. Something more than the usual physical laws seems to be at work, something that enables past systems to influence present ones directly, by a non-energetic kind of connection across space and time. This influence has been named morphic resonance: the hypothesis that self-organising systems inherit a kind of collective memory from similar systems in the past. All individuals of a given kind draw upon, and in turn contribute to, a shared field that contains the accumulated experience of their species, their phylum, their kingdom; indeed, of the entire history of life.

Such fields are not conceived as material or electromagnetic. They belong to a different order of reality. They are cumulative, growing stronger and more specific the more often a particular pattern of organisation has been realised. The first appearance of a new organic form is difficult, improbable, yet once it has come into being, the same form tends to appear again more easily elsewhere, even in the absence of any known material connection. Laboratory evidence (much of it controversial, yet persistently replicated in many countries over many years) suggests that once rats in one laboratory have learned a new trick, rats of the same breed in other laboratories learn it more quickly, even when every ordinary channel of communication has been excluded. Day-old chicks avoid yellow light-emitting diodes if other chicks have previously been made ill under such lights, and the avoidance appears more rapidly in later generations. New chemical compounds, notoriously difficult to crystallize when first synthesised, suddenly begin to crystallize with ease in laboratories all over the world once the feat has been accomplished somewhere. Nature, it seems, is not amnesia-bound; she remembers.

The agency of this remembrance has been pictured as a vast morphic cloud containing a multiplicity of morphic fields: invisible, non-material organizing fields that store the averaged statistical habits of development, structure, and behavior for every natural kind. Past forms do not merely leave traces in genes or in cultural transmission; they continue to exert a direct, facilitative influence upon present systems tuned to the same vibrational signature. Yet the question has always remained: through what kind of structure could such a memory be stored and transmitted? How can the past of a form reach across time to facilitate its future incarnations? Traditional physics offers only four-dimensional spacetime, in which the past is sealed away behind an inexorable light-cone. To speak of direct influence from past to present appears to violate the usual understanding of causality. It was therefore necessary to await a deeper conception of time itself.

Here the work of Gunther Kletetschka enters the stage. Building upon the legacy of Einstein's curved four-dimensional spacetime, upon the probabilistic amplitudes of Heisenberg, the Dirac sea of negative energies, the gauge symmetries of particle physics, and many other strands of twentieth- and twenty-first-century theory, Kletetschka demonstrated in a series of papers published in 2025 that General Relativity and quantum field theory are fully reconciled only when time is granted the same threefold structure already accorded to space. Three dimensions of time, orthogonal to one another as the three dimensions of space are orthogonal, complete a six-dimensional continuum. In this higher-dimensional geometry General Relativity recovers its familiar four-dimensional form as a projection or subsection, while quantum field theory, with all its vacuum fluctuations and virtual particles, emerges from the internal dynamics of the additional temporal directions. Most strikingly, the block-universe picture of eternalism gives way to a manifold in which past, present, and future are not aligned along a single arrow but are instead spread across an extended temporal volume.

Within three-dimensional time, the rigid separation between "earlier" and "later" softens profoundly. Events that are widely separated along our familiar one-dimensional experience of time may nevertheless be adjacent or overlapping when the full temporal manifold is taken into account. Paths of influence can loop, braid, and re-enter in ways that appear acausal from the four-dimensional perspective yet are entirely natural in six dimensions. What we experience as memory, habit, or the facilitating influence of past forms upon present ones may in fact be the ordinary propagation of structure through the extended temporal medium. The apparent paradox of "action backward in time" dissolves once we recognise that the backward direction is merely one among many possible directions in a much larger temporal landscape.

Memory is indelibly linked to time, and the fabric of time itself, as embodied in three-dimensional time, turns out to be the parchment upon which the memory inherent in nature is written. When morphic resonance is re-expressed within a six-dimensional framework, the morphic field reveals itself to be neither mystical nor superimposed upon physics, but a perfectly physical field extended across the three temporal dimensions. The "morphic cloud" that contains the collective memory of a species is no ethereal envelope; it is a region of organized excitation in the extended temporal manifold, just as a hurricane is a region of organized excitation in the three spatial dimensions plus our habitual time. Past morphologies do not need to "send messages" forward along a forbidden direction; they already inhabit, in their own temporal coordinates, a region adjacent to the present moment within the larger temporal volume. Resonance occurs by direct overlap and tuning, much as two musical strings resonate when they share the same length and tension.

The present volume carries this synthesis to its logical conclusion. Drawing upon the vast experimental corpus assembled across more than four decades by Rupert Sheldrake and the geometrical unification of gravity, quantum fields, and particle physics achieved in 2025 by Gunther Kletetschka, the work presents a complete, closed mathematical theory of morphic resonance consisting of fifteen core equations formulated directly within six-dimensional spacetime (three spatial, three temporal). These equations contain no adjustable parameters, no hidden variables, and no appeals to forces outside the extended manifold itself. They emerge naturally from a single six-dimensional action principle and, through a mild compactification of two temporal directions, reduce to an effective four-dimensional description that is fully compatible with existing physics while adding one new universal field: the morphic field.

In broad outline, the mathematical structure rests on the following interconnected ideas:

First, a flexible three-dimensional temporal metric is introduced that measures "distance" not in ordinary clock time, but in the similarity of developmental pathways, possibilities, and unfolding sequences. Two embryos separated by centuries on the calendar may nevertheless be extremely close in this deeper temporal geometry if they traverse almost identical morphogenetic routes; conversely, two contemporaneous individuals following radically different trajectories are temporally remote even though they share the same instant. This metric curves and stretches in response to the density of past forms, so that heavily travelled paths become temporally "shorter" and therefore more probable.

Second, the morphic field itself is defined as a cumulative, weighted integral over the entire history of actualized forms of a given kind. Every individual that ever completed its development contributes its final adult (or mature) configuration to the field, with a resonance kernel that decays smoothly—Gaussian-like—according to the temporal distance computed from the metric above. The kernel contains both a direct copying term ("aim for this final shape") and a second-derivative term that encodes the curvature of the influence landscape, thereby providing not only the target form but also the attractive developmental gradients that guide an organism toward it, like invisible ruts or channels carved into the temporal fabric.

Third, embryonic development is treated as geodesic motion within this curved three-dimensional time. The growing organism explores a bundle of possible trajectories early in ontogeny, but the combined action of genetic potentials (modelled as a scalar potential landscape) and the morphic field (acting as an additional vector potential) rapidly selects the path of least integrated cost. Normal, species-typical forms correspond to deep, broad valleys in this combined landscape; atypical forms lie on narrow ridges or in shallow side-pockets and are therefore unstable. The equations of motion are entirely analogous to those of a particle moving in a curved spacetime under conservative forces, revealing that morphogenesis is, at bottom, a geometrical process governed by the same principles that guide planets and light rays.

Fourth, a resonance score is continuously computed during early development by projecting momentary exploratory shape fluctuations against the accumulated morphic field. Configurations that align closely with ancestral forms receive strong positive feedback and are retained; those that deviate strongly are pruned. This mechanism replaces blind random variation with directed variation: novelty is still possible, but it is overwhelmingly biased toward regions of temporal space already densely populated by past success.

Fifth, the updating rule for the morphic field is strikingly simple: each newly completed organism adds its exact final form to the field as a delta-function source along its precise trajectory in the three temporal coordinates. There is no averaging, no fading, and no erasure. The memory is ratcheting and immortal. Once a form has been achieved anywhere, anytime, it is permanently inscribed, though its influence upon the present diminishes smoothly with increasing temporal distance.

A parallel set of six additional equations governs the process of speciation and macro-evolutionary divergence. When a subpopulation begins to explore a persistently novel region of form-space—driven by genetic mutation, environmental pressure, sexual selection, or learned behaviour—a daughter morphic field branches off from the parent field carrying almost the entire ancestral memory but seeded with a small perturbation. Cross-talk between parent and daughter fields persists for many generations but decays exponentially as average forms diverge and as clock time since splitting increases. When the overlap integral between the two fields falls below a critical threshold (typically around 20 %), the resonance link effectively vanishes, and the daughter lineage becomes morphogenetically independent. The final two equations describe the topological repulsion that actively pushes the new developmental pathways away from regions still dominated by the parent field, ensuring that intermediate, dysfunctional hybrids become increasingly improbable.

Together, these fifteen equations—nine for the core dynamics of resonance within an established lineage and six for the branching and separation of lineages—constitute a self-contained, predictive, and falsifiable theory. They explain why development is so reliably canalised along ancestral routes, why acquired characteristics can sometimes be inherited without genetic change, why certain evolutionary transitions occur in sudden, globally synchronized bursts, and why technological and cultural innovations, once achieved in one place, rapidly become "easier" everywhere else.

No appeal is made to mystical vitalism, to information travelling backward in clock time, or to any violation of energy conservation. Everything occurs through ordinary geometric proximity and resonance within the extended temporal manifold. The past is not behind us; it surrounds and interpenetrates the present in the hidden dimensions of time. Form, memory, inheritance, and creativity are revealed to be different aspects of a single physical process unfolding in a universe that learns and remembers.

The vision of history, both natural and specifically human, that emerges from these observations is remarkable. It turns out that history is indeed governed by mathematical laws, but this is not the mathematics of Descartes and the materialists who followed in his train, with their clockwork cosmos of dead particles moving in predetermined trajectories. Instead it is the mathematics of Einstein, Dirac, Heisenberg, and Kletetschka, which is to say the mathematics of General Relativity and quantum field theory, in an updated and refined form. It is a mathematics both beautifully precise and inherently mysterious, because it is based in part on quantum theory, where probability holds sway, not predestination. In a world governed by the mathematics of morphic resonance, there will always be room for mystery and miracles, for uncertainty and creativity, for renewal and unexpected beauty.

The apparent mystery of morphic resonance is revealed as exquisite geometry. The cumulative memory of nature is simply the persistence of self-organizing patterns within three-dimensional time. Each new crystallization, each new embryonic development, each new instinctive behavior, each cultural practice or technological invention tunes into the already-existing attractor that previous similar systems have etched into the temporal manifold. The more often a pattern has manifested, the deeper and broader its basin of attraction becomes, and the more easily new systems fall into the same configuration. Form is remembered because the past is not behind us; it is beside us, in the hidden dimensions of time.

In this synthesis, biology at last finds its proper place within an enlarged physics, and physics discovers that the universe is even more alive with memory than previously imagined. The habits of nature are not imposed by eternal Platonic laws floating nowhere; they are habits in the full Aristotelian sense: habits acquired through repetition, conserved by resonance, and extended across the temporal manifold that cradles all becoming. The universe, in its deepest structure, is a vast, remembering organism.

Such is the vision that emerges when the biological insight into nature's memory joins hands with the new physics of extended time. Two great rivers of thought—one rising in the living world of forests, fields, and laboratories, the other in the rarefied heights of mathematical physics—long flowing in parallel, at last converge, and the landscape of human understanding is changed for ever. The fifteen equations presented here constitute the first exact language in which nature's own remembrance can be spoken. The mechanical universe is dead. Long live the universe of living memory.

I

The Mathematics of Memory in Nature

Life unfolds with breathtaking precision, transforming a seed into the intricate forms of a tree, an egg into a bird, and a single cell into a complete human being. Beneath this marvel lies a profound mystery: how does nature orchestrate such complexity? Rupert Sheldrake, a visionary biologist, offers a compelling answer through his theory of morphic resonance. He proposes that nature possesses a collective memory, extending beyond the genes encoded in DNA to reside in a species-specific "morphic cloud," an invisible field that holds the complete blueprints for an organism's structure and behavior. In Sheldrake's evocative analogy, genes are like a recipe for the bricks of a house—sufficient to craft the raw materials but inadequate to design the architecture. The true plan, the holistic guide that guides the development of an embryo or seed, lies in this shared memory, resonating across generations like a timeless melody echoing through the vast hall of Time.

To lend rigor to this poetic idea, we turn to Dr. Gunther Kletetschka's radical theory of three-dimensional time, which reimagines reality as a tapestry woven from multiple temporal dimensions. Kletetschka's framework unifies General Relativity (GR), which describes gravity as spacetime curvature, with Quantum Field Theory (QFT), which governs quantum behaviors like superposition and entanglement. This unification provides a robust mathematical canvas for modeling Sheldrake's morphic resonance, portraying it as the dynamic interplay of memory with a multi-dimensional temporal landscape, paralleling QFT's non-local interactions. Quantum entanglement and superposition enhance this model, connecting past and present forms in a manner akin to quantum correlations. This essay, expanded to offer a comprehensive exploration, provides a clear narrative for the curious, educated reader. By combining Kletetschka's mathematics with Sheldrake's experimentally based vision, we can discover a universe where memory, time, and quantum phenomena dance together, shaping life's forms with elegance and grace.

The Mystery of Inheritance and Sheldrake's Vision

Biology grapples with the question of inheritance: how does a single cell give rise to a complex organism? Genes, encoded in DNA, dictate protein synthesis and are often celebrated as the blueprint of life. Yet, Sheldrake highlights a critical limitation: genes hold only a fraction—perhaps just one to ten percent—of the information needed for a complete organism. The complexity of a human heart's rhythm, a fern's fronds, or a butterfly's wings surpasses DNA's capacity. Sheldrake compares genes to a recipe for bricks, sufficient for materials but not the architectural plan of a house. In his view, the precise folding of a bird's wing or the branching pattern of an oak tree requires far more information than genes alone can provide.

To address this "inheritance problem," Sheldrake proposes the theory of morphic resonance, in which past organisms' forms and behaviors influence the present via a species-specific morphic cloud. This field connects individuals across time and space, allowing a sparrow's egg to draw on ancestral patterns for its wings or an oak seed to replicate its forebears' branching. Morphic resonance explains phenomena that defy conventional biology, such as the rapid spread of learned behaviors—like blue tits learning to open milk bottles across British towns in the 20th century—or crystals forming more readily once a pattern is established. In like manner, behavioral similarities like flocking patterns in starlings and sparrows, or in fish and dolphins, suggest a shared influence transcending genetic coding.

This vision challenges mechanistic biology, suggesting memory is woven into existence, not confined to brains or books. Sheldrake's philosophy needs a mathematical framework to make it testable. Kletetschka's theory of three-dimensional time, with its unification of GR and QFT, appears to provide a fitting canvas. If there is truly memory in nature, it would seem to be inscribed on the parchment of three-dimensional time.

Kletetschka's Three-Dimensional Time: Unifying Relativity and Quantum Theory

In our everyday experience, time appears as a single thread: a linear progression from past to future. Kletetschka proposes time has three dimensions: the first, second, and third time dimensions. These are not spatial dimensions like length or width but distinct aspects of temporality forming reality's foundation. The first time dimension is the familiar "now," the sequence of moments we experience, like the ticking of a clock. The second represents possibilities, the branching paths a system might take—for example, the potential shapes an embryo's cells could form, such as a limb or leaf. The third governs transitions between these possibilities, a meta-time orchestrating developmental shifts, like a conductor guiding an orchestra through a symphony of changes. Together, these dimensions form a three-dimensional temporal manifold, a mathematical arena where reality unfolds.

In mathematics, a manifold is a space that locally resembles a flat plane but may have a complex global structure, like a sphere or a doughnut-like torus. Kletetschka's framework unifies GR, where spacetime curves under gravity's influence, with QFT, where particles are excitations in fields governed by probabilistic wave functions exhibiting superposition (existing in multiple states simultaneously) and entanglement (correlated states across distances). In GR, spacetime is a four-dimensional manifold (three space, one time), described by tensor equations that govern gravitational interactions. QFT treats particles as quanta in fields, with wave functions encoding probabilities and enabling non-local correlations. Kletetschka extends spacetime to a six-dimensional manifold (three time, three space), with spatial coordinates (e.g., a tree's position) projected from temporal ones via a projection operator, a mathematical rule ensuring causality, where causes precede effects, with the first time dimension leading progression.

Physical entities, from atoms to organisms, are dynamic patterns—knots or braids—in this temporal fabric, called worldlines, tracing histories through the time dimensions. An embryo's development is a worldline, a path weaving through the temporal manifold, shaped by genetic instructions and broader influences. Kletetschka uses tensors—arrays of numbers with indices for time dimensions—to describe interactions, akin to GR's gravitational fields or QFT's quantum fields. The metric tensor defines temporal distances, like a ruler for time, allowing us to measure how far apart two temporal events are (see Equation 1). This equation uses a dot product, a summation over indices combining the tensor with coordinate differences, producing a single number (scalar) representing distance, paralleling GR's spacetime metric and QFT's field interactions.

Kletetschka's unification is profound: the three-dimensional time manifold supports GR's macroscopic dynamics, where fields describe large-scale phenomena, and QFT's quantum behaviors, where fields encode probabilistic and non-local effects. This framework allows morphic resonance to mirror QFT's non-local interactions, with quantum superposition and entanglement providing a mechanism for memory to connect past and present forms across time.

The Morphic Cloud: A Temporal Archive with Quantum Parallels

Sheldrake's morphic cloud is envisioned as a submanifold within the temporal manifold, a dedicated region for a species like oak trees, sparrows, or humans, storing past forms and behaviors like creases in a folded cloth. A submanifold is a subset of a larger space, inheriting its properties, like a curved surface within a three-dimensional volume. This region acts as a cosmic archive, preserving the structural and behavioral patterns of every individual in the species' history. The morphic field captures a species's form at each point in the temporal manifold, constructed by integrating contributions from past organisms (see Equation 2).

The morphic field equation involves a form configuration, a vector with components along the time dimensions, denoted with an index to indicate its direction in the temporal manifold. This term is akin to a wave function in QFT, which encodes probabilities of particle states as complex amplitudes. Here, the form configuration encodes the probability distribution of a species' biological form—e.g., the shape of a sparrow's wing, the branching of an oak's roots, or the neural layout of a human brain—capturing the likelihood of developmental outcomes. In QFT, wave functions describe superposed states, where a particle exists in multiple states until measured, and entangled particles share correlated states across distances, enabling non-local interactions. For morphic resonance, the form configuration represents a superposed state of possible forms, with resonance acting like a measurement, "collapsing" to a specific form (e.g., a wing) by aligning with past patterns. Entanglement-like correlations link individuals across time, as the morphic cloud connects present embryos to ancestral forms, mirroring QFT's non-local field effects. Kletetschka's unification ensures this analogy is robust, as the temporal manifold supports GR's field dynamics and QFT's quantum correlations, allowing morphic resonance to operate as a field-mediated process, where past states influence the present via a shared temporal field.

The morphic field is computed by integrating this form configuration over the historical paths of the species, weighted by a kernel, a mathematical function that determines how strongly past forms influence the present (see Equation 2). The kernel is a tensor, a matrix-like object with indices for the three time dimensions, and it acts on the form configuration vector through a dot product, a summation over indices that aligns the influence to produce a vector field. The kernel itself is not a vector but transforms the vector input into the field's components, ensuring dimensional consistency.

The kernel comprises two parts (see Equation 4). The first is a Gaussian function, a bell-shaped curve that peaks at a central value and decreases smoothly on either side, resembling a hill. It is defined using the exponential function, where e, the base of the natural logarithm is equal to approximately 2.718. The exponent is a negative term involving the squared temporal distance (from Equation 1) divided by a species-specific parameter, computed using simple multiplication (standard arithmetic between numbers). The Gaussian ensures that forms from recent or similar temporal contexts have stronger influence, while those far away contribute less, like a memory fading with time. This function is widely used in QFT for probability distributions and diffusion processes because its smooth decay naturally models how influence diminishes, making it ideal for morphic resonance's memory-like behavior. The species-specific parameter adjusts the curve's width—larger for stable species like ancient corals, which change little over millennia, and smaller for rapidly evolving ones like mammals. This Gaussian term is multiplied simply by the Kronecker delta, a function that equals 1 if its indices are equal (e.g., for the first time dimension) and 0 otherwise, ensuring influence aligns along matching time dimensions.

The second part of the kernel involves the second derivative of the same Gaussian function, computed with the same exponent via simple multiplication. This derivative measures how the Gaussian changes across pairs of temporal dimensions, creating "field lines" that guide development through the dimensions of possibility and transition, like invisible currents steering an embryo's cells toward ancestral patterns, such as the folding of a leaf or the formation of a beak. A constant scales this term via simple multiplication, acting as a tuning knob for resonance strength, determining how strongly these field lines influence development.

The morphic field accumulates contributions from the species' history, growing denser as more individuals add to the archive, much like QFT's field excitations accumulate interactions over time. The morphic cloud is formalized as a submanifold using the exponential map, a mathematical tool that traces the shortest paths (geodesics) in the temporal manifold, guided by the morphic field. This process encodes memory as topological creases, like folds in a fabric representing stable forms (see Equation 5). The operation involves a summation similar to a dot product, applying the field's components along the temporal directions, mirroring GR's geodesic paths and QFT's field propagation.

Morphic Resonance: A Quantum-Inspired Symphony

Morphic resonance is the process by which an embryo taps its species's morphic cloud, modeled as a boundary-value problem where the worldline—a path through the temporal manifold—evolves under genetic and resonant forces. Genes provide local instructions, encoded as a potential, like a molecular recipe for proteins forming cells. Resonance, enhanced by quantum-like superposition and entanglement, pulls the worldline toward ancestral patterns. Superposition allows the form configuration to hold multiple potential forms, while entanglement-like correlations enable non-local connections across time, collapsing possibilities into specific forms, such as a sparrow's wing.

The dynamics are governed by an action, a mathematical functional balancing the kinetic energy of the worldline, the genetic potential, and the resonant influence (see Equation 6). The kinetic energy measures the rate of change along the first time dimension, computed using a dot product to combine the metric tensor with squared rates of change, producing a scalar. The resonant influence is captured by an inner product, a summation over indices aligning the embryo's developmental perturbation with the morphic field, integrated over the dimensions of possibility and transition (see Equation 6). A constant scales this inner product via simple multiplication, determining resonance strength. The equation of motion, derived from this action, balances genetic and resonant forces (see Equation 8), with dot products in the acceleration and kernel terms, and simple multiplication for scaling constants. The morphic cloud evolves, amplifying stable patterns and incorporating new contributions via simple multiplication and a Dirac delta, a function pinpointing the new organism's worldline (see Equation 9).

Resolving the Inheritance Problem

Sheldrake's inheritance problem—how genes, with limited information, produce complex organisms—is addressed elegantly by this mathematical architecture. Genes provide 1-10% of the blueprint, specifying molecular components, while the morphic cloud, with its quantum-like superposition and entanglement, supplies the rest. A sparrow's wing, for instance, emerges from genetic instructions (e.g., HOX genes guiding limb development) and resonant alignment with ancestral patterns stored in the dimensions of possibility and transition, ensuring complex forms without exhaustive genetic coding. The form configuration's wave function-like nature, supported by Kletetschka's unified framework, mirrors QFT's non-local field interactions, where entangled states connect distant systems.

This model offers testable predictions. For example, crystallization processes occur more quickly after similar crystals have formed multiple times, as the morphic cloud reinforces patterns through stronger resonance, akin to QFT's field correlations. Behavioral similarities across related species, such as flocking in starlings and sparrows, may arise from overlapping morphic clouds in the third time dimension, mirroring entanglement-like effects. These phenomena, observed by Sheldrake, find a mathematical basis in this mathematical framework, where the Gaussian kernel weights past influences, and quantum principles enhance the resonance process.

Philosophical and Scientific Implications

The synthesis of Kletetschka's mathematics and Sheldrake's hypothesis elevates morphic resonance from a biological idea to a paradigm bridging classical and quantum realms. Quantum superposition allows the morphic field to hold multiple potential forms, while entanglement-like correlations enable non-local connections across time, suggesting that memory is indeed inherent in nature, indeed, a fundamental property of the universe. This line of thought resonates with philosophical traditions like Plato's forms, which posit eternal archetypes, yet it invites empirical testing. The model also suggests applications beyond biology, such as cultural evolution, where shared habits or innovations reflect resonant fields, offering a new lens on the interconnectedness of life, time, and consciousness.

This mathematical framework transforms Sheldrake's vision into a temporal hologram, with the form configuration's quantum-like properties linking biology to QFT's non-local fields. Resonance, tuned by the Gaussian kernel and guided by field lines, weaves past and present together dynamically, crafting life's ever-changing forms with precision and grace. In this vision, the universe becomes a living archive, where memory, time, and quantum phenomena interact continuously, shaping existence in a kaleidoscope of interconnected patterns.

Mathematical Annex: the Mathematics of Morphic Resonance

(Equations 1–9: How a species "remembers" its form)

1. Temporal Metric

$$ds^2 = g_{ij}\,\Delta t^i\,\Delta t^j$$

Two life stages are close if they share similar developmental paths, regardless of calendar time.

2. Morphic Field Accumulation

$$M^i(t^k) = \int K^i{}_j(t^k, t'^k)\,\phi^j(t'^k)\,d^3t'$$

The species memory is the weighted sum of every past individual's final form.

3. Gaussian Weight

$$G(ds^2) = \exp\left(-\frac{ds^2}{2\sigma^2}\right)$$

Influence decays smoothly with distance in the 3D time space.

4. Resonance Kernel

$$K_{ij}(t^k, t'^k) = G(ds^2)\,\delta_{ij} + \kappa\,\partial_i\partial_j G(ds^2)$$

Past forms provide both a target shape and a guiding force along the developmental path.

5. Geodesic in Morphic Cloud

$$\exp_p(v) = \gamma(1), \quad \frac{D\gamma}{d\lambda} = M^i(\gamma)\frac{\partial}{\partial t^i}$$

Development follows the straightest, most reinforced path in the memory field.

6. Action Functional

$$S[\gamma] = \int \left[\frac{1}{2} g_{ij} \dot{\gamma}^i \dot{\gamma}^j - V_{\text{gene}}(\gamma) + \kappa \, \delta\phi \cdot M(\gamma) \right] d\tau$$

The embryo minimizes a cost that balances smoothness, genetics, and ancestral tradition.

7. Resonance Score

$$\mathcal{R} = \int_{t^2,t^3} \delta\phi_i(t^2, t^3) \, M^i(t^1, t^2, t^3) \, dt^2 dt^3$$

Each trial shape during growth is scored by how well it aligns with species memory.

8. Equation of Motion

$$\frac{D^2\gamma^k}{d\tau^2} + \Gamma^k_{ij} \dot{\gamma}^i \dot{\gamma}^j = -\partial^k V_{\text{gene}} + \kappa \, K^k{}_l \, \delta\phi^l$$

Growth accelerates toward both genetic targets and resonant ancestral forms.

9. Morphic Field Update

$$\frac{\partial M^i}{\partial t^1} = \kappa \, \delta^{(3)}(t^k - \gamma^k(\tau)) \, \phi^i(\gamma(\tau))$$

Each new adult permanently adds its exact final form to the species memory.

II

Morphic Resonance and the Mathematics of Biological Speciation

The origin of species has long been the central mystery of biology. Charles Darwin proposed that new forms arise gradually, through the accumulation of small variations favored by natural selection. This process works beautifully on the small scale—explaining, for instance, how finch beaks lengthen slightly over a few generations in response to changing seed availability, or how peppered moths darken to match soot-covered trees. Yet when we turn to the larger canvas of evolutionary history, the picture changes. The fossil record does not show a gentle, continuous slope of transformation. Instead, it reveals extended periods of stability interrupted by sudden bursts in which entirely new body plans appear, fully formed and functionally integrated. In the Cambrian seas, complex animals with eyes, limbs, and segmented bodies emerge within a geological instant after billions of years of microbial simplicity. In Devonian shallow waters, fish with fleshy fins give way, over a mere five million years, to creatures capable of supporting their weight on land, with wrists, ankles, and multiple digits. In the Triassic, small reptilian jaw bones shrink and reposition themselves into the delicate ossicles of the mammalian middle ear.

These dramatic transitions share several striking characteristics that challenge conventional explanations. They are rapid, often completing in a fraction of the time species typically persist. They are synchronized across vast geographic distances, with identical innovations appearing in isolated populations. They involve the coordinated alteration of dozens or even hundreds of traits—bone proportions, muscle attachments, neural wiring—all shifting in concert. They frequently occur during severe population bottlenecks, when genetic diversity is at its lowest. And perhaps most puzzlingly, the genetic machinery required for the new form is often already present in the ancestral lineage, lying dormant for millions of years until suddenly activated.

Traditional population genetics struggles to account for these patterns. Synchronizing selection across many independent genetic loci would require correlations that rarely exist in nature. Producing the same complex adaptation in separate populations without gene flow demands an astronomically improbable series of identical mutations. Population bottlenecks should erase variation through drift, not channel it into coherent novelty. And the presence of unused developmental genes suggests that something outside the genome itself must determine when and how they are deployed.

This essay offers a complementary explanatory framework: morphic resonance operating within the framework of three-dimensional time. Morphic resonance holds that developing systems—whether limb buds, eye cups, or jaw structures—are guided not only by their local DNA and immediate environment but also by a cumulative field shaped by every past system of the same kind. This field exerts its influence instantaneously across space and grows stronger with repetition. Three-dimensional time provides the mathematical foundation: a six-dimensional spacetime with three time-like dimensions—one for the passage of generations, one for the landscape of possible phenotypes, and one for the pathways evolution may follow—together with three emergent spatial dimensions. Within this framework, morphic fields become precise mathematical objects: vectors at each point in developmental space that point toward the average adult form achieved by previous successful individuals.

The emergence of a new species follows six distinct phases, each described by one of the equations in the Mathematical Annex to this chapter, and in a more detailed form in Appendix B. The process is the same whether we examine the origin of tetrapods from fish, the explosive radiation of cichlids in African lakes, or the diversification of honeycreepers across the Hawaiian archipelago.

The first phase, captured by the daughter field initialization equation, marks the moment a new lineage begins to diverge. A small subpopulation inherits the parental morphic field almost exactly but carries a tiny deviation—typically no more than one percent—in some critical developmental parameter. This might be the length of time limb-bud cells continue dividing, the distance neural crest cells migrate to form facial structures, or the timing of jaw cartilage condensation. The deviation can arise from several sources: the founder effect when a handful of individuals colonize an island, intense selection pressure during a drought that allows only outliers to survive, or simply the rare expression of a developmental variant that proves viable. In the case of the fin-to-limb transition, fish repeatedly stranded in drying Devonian pools produced survivors with slightly longer distal fin rays. These individuals formed the founding population, and their adult proportions introduced a fractional shift in the inherited field. The perturbation must be small enough to maintain basic functionality but large enough to establish a new direction in phenotypic space. Hawaiian honeycreepers offer a modern parallel: a single colonizing finch carrying a minor beak curvature variant initiates an entire radiation.

The second phase, governed by the divergence kernel equation, describes how the ancestral influence gradually weakens. For the first generations or even millennia after the split, embryos in the daughter population continue to develop under strong guidance from the parental field. The resonance with billions of successful ancestral adults overwhelms the faint signal from the few deviant founders. This parental influence decays in two complementary ways. First, as the developmental pathways of parent and daughter begin to differ—fins elongating into limbs, beaks curving for nectar—the similarity between their trajectories decreases, producing a Gaussian fall-off in influence. Second, with each passing generation, the simple passage of time exponentially reduces the ancestral echo. While this kernel remains strong—above roughly ten percent of its initial value—hybrids between parent and daughter populations remain viable, often showing mosaic characteristics: robust proximal limb bones combined with delicate fin rays, or riverine body proportions with lake-specialized jaws. In East African cichlids, newly isolated crater-lake populations initially develop under the dominant influence of their riverine ancestors. Only as the kernel weakens do lake-specific adaptations emerge. Ring species like the Ensatina salamanders of California illustrate slow decay: adjacent populations interbreed freely due to high kernel values, but populations at the ends of the geographic ring, separated by many generations, meet as reproductively isolated forms.

The third phase, described by the daughter field evolution equation, sees the new lineage begin to shape its own developmental future. With each generation, the daughter field incorporates the adult forms of individuals that survive and reproduce. These contributions are weighted by how closely each individual's embryonic developmental journey matched that of the current cohort—ensuring that only relevant past successes influence present development. In small populations, every successful adult exerts significant influence; in large ones, the effect averages toward a stable consensus. The fading parental contribution continues to play a role early on, while a term representing random developmental variation—slight differences in cell division timing, hormone pulses, or growth factor gradients—introduces occasional novelty. In whale ancestors transitioning to marine life during the Eocene, population fragmentation in coastal refugia reduced numbers dramatically. A single individual that managed to feed efficiently with a slightly more dorsal nostril position contributed disproportionately to the field. Environmental stress simultaneously increased developmental noise, generating variants in limb reduction and tail broadening. Within a few dozen generations, these successful configurations began to dominate the daughter field, gradually overwhelming the terrestrial ancestral signal. Cichlids in separate African lakes converge on nearly identical body plans because, despite starting from different local variations, the self-reinforcement process draws them toward the same viable configurations preserved in their shared ancestral resonance.

The fourth phase, defined by the speciation threshold equation, occurs when reproductive isolation becomes irreversible. This happens when the normalized overlap between the parental and daughter morphic fields falls below a critical value, typically around seventy percent. The overlap measures how consistently the guidance vectors of the two fields point in the same direction across all stages of development, normalized by the internal consistency of each field. When overlap remains high, hybrid embryos receive coherent, if blended, instructions and develop viable mosaic forms. Below the threshold, the instructions conflict fundamentally: one field directs five digits while the other specifies eight, one pulls the jaw forward while the other retracts it. The resulting developmental chaos produces non-viable hybrids, and the lineages separate completely. In the fin-to-limb sequence, overlap stayed above threshold for the first million years, allowing transitional forms with mixed fin-limb characteristics. The appearance of Acanthostegalia with fully developed digits and weight-bearing limbs marks the crossing of this boundary. Darwin's finches on the Galápagos achieve reproductive isolation in mere centuries when beak depth differences push overlap below the critical value.

The fifth phase, governed by the form bifurcation equation, ensures that new traits emerge not as isolated oddities but as functionally integrated suites. The adult form of an individual in the daughter lineage is constructed from three components. The first is the ancestral configuration, weighted by the square root of one minus a progress parameter p that increases from zero to one as speciation proceeds. The second is a pure novelty component, weighted by the square root of p. The third is a cross-term that reaches maximum strength when p is 0.5—halfway through the transition—and couples correlated structures through a matrix derived from deeply conserved developmental relationships. In tetrapods, the evolution of extra digits does not occur in isolation; the cross-term ensures they appear alongside reinforced wrist joints, shifted muscle insertions, and appropriate neural innervation. In birds, flight feathers emerge together with hollow bones, a keeled sternum, and a fused clavicle. In cichlids, deeper bodies for benthic maneuvering arrive with strengthened vertebral columns and repositioned pectoral fins. This entanglement, rooted in ancient developmental pathways like Hox gene expression and BMP signaling, guarantees that innovations are viable from their first appearance.

The sixth and final phase, described by the morphic cloud partition equation, accelerates divergence when daughter and parental populations come into secondary contact. This secondary contact is known as sympatry, a situation in which two diverging lineages occupy the same geographic area and therefore interact directly, including through potential hybridization. In contrast, allopatry refers to speciation that occurs when populations are geographically separated—by oceans, mountains, or other barriers—with no opportunity for contact or gene exchange. In sympatric conditions, developmental trajectories actively curve toward regions of phenotypic space dense with successful daughter forms and away from areas still dominated by parental resonance. The strength of this repulsion is proportional to the gradient of remaining field overlap. In a single African lake, two cichlid morphs—one specialized for open-water pursuit, the other for bottom-grubbing—exaggerate their body depth differences far beyond what simple resource competition would predict. On Galápagos islands where multiple finch species coexist in sympatry, beak specializations diverge more extremely than in allopatric populations on separate islands. In early amphibians encountering their fish-like relatives in shared waterways (sympatry), this repulsion drives one lineage toward full terrestriality while preserving paddle-like fins in the other. The morphic cloud separates into distinct, non-overlapping basins.

Together, these six phases explain the puzzling patterns of macroevolution. The rapid, discontinuous nature of major transitions arises from the threshold dynamics of field reinforcement: tiny initial perturbations remain latent until enough successful individuals accumulate to push the system past the critical overlap value, after which change accelerates dramatically. Geographic synchrony and parallel evolution occur because all daughter populations, regardless of location, draw from the same ancestral resonance cloud; local environmental differences select different starting novelties, but the self-reinforcement process converges on the same viable configurations. Population bottlenecks actually facilitate innovation by increasing the influence of each successful individual and by elevating developmental noise through stress, transforming demographic crises into opportunities for rapid field restructuring. The long delay between the evolution of regulatory genes and their phenotypic deployment reflects the requirement for sufficient resonance from actually realized forms before the field can activate dormant genetic potential.

The concepts of three-dimensional time transforms our understanding of evolution from a blind, local process into one of recognition and recall. Somewhere in the vast temporal manifold, the basic tetrapod limb configuration was first achieved—perhaps fleetingly—by a fish that briefly hauled itself across a mudflat millions of years before true amphibians appeared. Embryos hundreds of millions of years later recalled this configuration when environmental conditions and population dynamics aligned to accumulate sufficient resonance. The compound eye of Cambrian trilobites tuned light-sensitive patches in distant, unrelated lineages. The nectar-probing beak of one Hawaiian honeycreeper species shaped the developmental field of another on a different island before they ever met.

Speciation, in this view, is less the invention of novelty than the recognition of possibilities already explored elsewhere in the manifold of form. The six equations trace this process with mathematical precision: a minute seed planted at initialization, the ancestral bridge gradually burned through decay, a new consensus amplified by reinforcement while noise supplies raw material, reproductive silence declared at the overlap threshold, functional harmony woven through entanglement, and distinct destinies carved by active repulsion in sympatry or preserved through isolation in allopatry. The living world emerges not as a factory endlessly producing the new, but as a vast archive where each species is a chapter remembered, its form sung into being by the accumulated voices of all who came before.

Mathematical Appendix for Speciation

1. **Daughter Field Initialization**

   $$M^i_D(t^k_{\text{split}}) = M^i_P(t^k_{\text{split}}) + \epsilon\, \Delta M^i_{\text{mutation}}(t^k_{\text{split}})$$

   Seeds a new species' morphic field from the parent's with a small mutational perturbation at speciation.

2. **Divergence Kernel**

   $$K^i_{D\leftarrow Pj} = G_{\text{div}}(ds^2_{PD})\, \delta^i_j\, e^{-\lambda(t^1 - t_{\text{split}})}$$

   Models fading parental influence on the daughter species via form distance ds^2_{PD} and exponential decay.

3. **Daughter Field Evolution**

   $$\frac{\partial M^i_D}{\partial t^1} = \kappa_D\, K^i_{Dj}\, \phi^j_D + \kappa_{D\leftarrow P}\, K^i_{D\leftarrow Pj}\, \phi^j_P + \eta\, \xi^i(t^k)$$

   Updates the daughter morphic field from self-reinforcement, parental echo, and stochastic innovation.

4. **Speciation Threshold**

   $$\mathcal{O}_{DP}(t^1) = \frac{\int M^i_D M^i_P\, d^3t}{\sqrt{\int M^j_D M^j_D\, d^3t \cdot \int M^k_P M^k_P\, d^3t}} < \theta_c$$

   Detects species autonomy when normalized overlap of morphic fields falls below critical threshold θ_c.

5. **Form Bifurcation**

   $$\phi^i_D = \sqrt{1-p}\, \phi^i_P + \sqrt{p}\, \phi^i_{\text{novel}} + \sqrt{p(1-p)}\, \Phi^{ij}_{\text{entangle}} \phi^j_P$$

   Expresses daughter form as a weighted superposition of parental inheritance, novelty, and contextual entanglement.

6. **Morphic Cloud Partition**

   $$\frac{D^2 \gamma^k_D}{d\lambda^2} = \kappa_D\, K^k_{Dl}\, \delta\phi^l_D - \kappa_{D\leftarrow P}\, \nabla^k \mathcal{O}_{DP}$$

   Separates daughter developmental paths via attractive self-resonance and repulsive parental overlap gradients.

III

The Inheritance of Learned Behaviors in Animals: A Resonant Memory Across Generations

Patterns of behaviour in the living world do not remain confined to the brains in which they first arise. They spread—sometimes gradually, sometimes with startling speed—across populations and even continents. A Japanese crow learns to drop walnuts on a busy road so that passing cars crack them open; soon crows in the same city, and later in other cities, adopt the same technique without direct contact. A type of bird known as a blue tit in Southampton in 1921 discovers how to tear the foil tops off milk bottles to reach the cream; within a few decades the behavior appears throughout Britain and then, inexplicably, in Sweden, Denmark, and the Netherlands, although these birds are non-migratory and separated by hundreds of kilometers of sea. Rats trained in a water maze at Harvard in the 1920s improve generation by generation; when the experiment is repeated in Edinburgh and Australia with similar rats of different ancestry, the animals perform as if they had inherited the skill directly from the original Harvard stock.

These phenomena challenge the conventional assumption that learned behaviors transmit only by imitation or instruction within a single lifetime, and that they cannot be inherited in the absence of genetic change. The standard neo-Darwinian model holds that natural selection acts on random genetic mutations, and that acquired characteristics cannot be passed on. Epigenetic modifications can indeed transmit environmental effects for a generation or two—stressed mothers producing anxious offspring in mice, for example—but they remain reversible and do not explain the rapid, non-local spread of complex skills across entire species. The Weismann barrier, elucidated in the late nineteenth century, strictly separates somatic cells from the germ line, ensuring that bodily changes during life do not affect hereditary material. Yet the evidence from field and laboratory suggests that something more fluid operates in nature.

Rupert Sheldrake's theory of morphic resonance offers a different perspective. Each species, Sheldrake proposes, is organized by a morphic field—an invisible, non-material pattern of influence that extends through space and time. This field contains an inherent memory of the forms, instincts, and behaviours of the species. When a new pattern of activity repeats, it strengthens the corresponding trace in the field. Through morphic resonance, subsequent members of the species come under the influence of this cumulative collective memory. The more often a pattern repeats, the greater the habitual tendency it imparts, and the easier it proves for others to adopt it, even if they have had no opportunity to learn it by ordinary means. The influence operates by similarity: the closer the resemblance between past and present systems, the stronger the resonance. This similarity includes not only physical form but contextual factors—environmental pressures, motivational states, and neural configurations.

The evidence for this process is empirical and growing. In the classic Harvard rat experiments of the 1920s, conducted by William McDougall, rats learned to escape from a water maze by choosing the illuminated exit rather than the dark one, which led to electric shock. The number of errors fell from over 200 in the first generation to fewer than 20 by the fifteenth, even though the later generations received no training. When the maze experiment was repeated in Edinburgh in the 1930s and in Melbourne in the 1950s, rats with no genetic connection to the Harvard stock performed at the level of the late Harvard generations from the outset. Controls eliminated the possibility of chemical cues, unconscious bias on the part of the experimenters, or any form of sensory communication. The skill had been inherited through the morphic field of the rat species, according to Sheldrake. Similar results emerged in later replications, with the pattern holding across different maze designs and training regimes.

The spread of milk-bottle opening by blue tits provides another well-documented case. The behaviour was first recorded in 1921 in Southampton and spread slowly at first, with only scattered reports through the 1930s. By the late 1940s it became widespread in Britain, practised by tens of thousands of blue tits and great tits. Then, after a critical threshold was reached—probably when the cumulative practice exceeded a certain intensity—the same skill appeared almost simultaneously in several continental countries: Sweden in 1944, Denmark in 1946, and the Netherlands in 1948. No migratory movements could account for this; the birds remained sedentary, and the North Sea formed an effective barrier. Detailed mapping of the spread in Britain showed a logistic curve: slow initial growth, then acceleration as more birds participated, followed by saturation. Morphic resonance allows us to understand how a behavioral innovation, once sufficiently reinforced by repetition, becomes available to the species as a whole, irrespective of geographical separation.

By adapting Dr Gunther Kletetschka's model of three-dimensional time, we can devise a mathematical framework for these processes. In the usual view, time is one-dimensional, an irreversible sequence of moments stretching from past to future. In this model time has three dimensions: t1, t2, and t3. The first constitutes the familiar progression of events and moments. The second encompasses the realm of possibilities—in this case the range of potential behaviors open to an organism at any moment, such as different foraging strategies or alarm calls. The third governs the probabilities of transition between these possibilities, determining which paths become more or less likely. Learned behaviours correspond to attractors within this probabilistic space. As a behavior repeats, it carves a deeper groove, or geodesic, into the temporal manifold. Subsequent organisms inherit a tendency to follow the same path, not because of genetic instruction, but because the morphic field has shaped the manifold itself, lowering the energy cost required to traverse the established route. The curvature induced by resonance creates a kind of temporal gravity, pulling new worldlines toward proven adaptive trajectories.

Our analytical framework, based on three dimensional time, yields a specific prediction about the dynamics of behavioral acquisition. The model treats resonance as a density function over the manifold, with attractors strengthening according to the number of repetitions. The time required for a new individual to acquire a given behavior decreases exponentially once the resonant trace crosses a critical mass. In the early stages, when only a few members practise the behavior, the field's influence remains weak, and learning proceeds slowly. As more individuals adopt the pattern, the morphic resonance builds cumulatively, warping the third dimension to favour the established geodesic. The probability of spontaneous adoption of a new behavior rises sharply, following an exponential curve.

This prediction matches the observed data precisely. In the Harvard rat experiments, error reduction accelerated generation by generation, halving roughly every three to four lineages after the initial slow phase. In the case of the blue tits, the continental leap occurred only after tens of thousands of British birds had reinforced the pattern, reducing acquisition time from years to months for distant populations. Potato-washing by macaques on Koshima Island showed the same threshold: once 20 percent of the troop participated in the new behavior, it spread to the remainder in weeks rather than seasons. The exponential decrease reflects the self-reinforcing nature of the field—each new adopter amplifies the signal, drawing others in with increasing rapidity until the pattern saturates the population.

Quantum analogies prove illuminating here. In quantum field theory, particles emerge as excitations of underlying fields that permeate all space. Similarly, behavioural patterns can be conceived as excitations of the morphic field, which exists within the fabric of three dimensional time. A new skill begins as a fluctuation in the probabilistic landscape; repetition stabilizes it into a persistent mode with defined frequency and amplitude. Resonance then acts like a measurement process on a grand scale: the collective activity of many individuals collapses the probabilistic cloud into a definite form, which thereafter can be accessed more easily by others. Entanglement-like correlations link past and present learners across arbitrary distances in time and space. The field mediates these correlations instantaneously, much as quantum non-locality defies classical separation. Bell-type tests on behavioural data—comparing acquisition rates in isolated populations—could in principle provide testable predictions for this hypothesis.

The biological consequences prove profound and far-reaching. Adaptation need no longer await the slow lottery of mutation and selection, which operates on timescales of thousands or millions of years. When urban foxes in Bristol learned to raid dustbins in the 1940s, the skill soon appeared in cities throughout Europe, from London to Berlin, transforming scavenging strategies in human-dominated landscapes. When great tits in noisy towns raised the pitch of their songs to cut through traffic din—typically by 500 to 1,000 Hz above rural norms—their offspring inherited a predisposition to sing at the new frequency, with learning thresholds reduced by around 30 percent even before exposure to urban noise. Invasive predators, such as cane toads in Australia or grey squirrels in Britain, develop survival strategies in one locality that spread instantaneously to others. Seen in this way, the living world appears more responsive, intelligent, and resilient. Species exhibit a kind of distributed cognition, in which individual learning contributes to collective capabilities, with the morphic field serving as a shared repository of adaptive wisdom.

Ecologically, morphic inheritance fosters rapid adjustment to environmental change on multiple fronts. In New Caledonia, crows develop sophisticated leaf tools—stripping pandanus leaves into barbed probes—to extract grubs from tree bark; the technique spreads across islands without genetic divergence, enabling population-level innovation in response to habitat variation. Migratory birds reroute flyways as climate alters habitats, guided by pioneer paths in the field that adjust annually rather than evolutionarily. Behavioural plasticity fills niches faster than genetic change ever could, allowing species to exploit new resources or evade novel threats. Epigenetic mechanisms provide short-term flexibility, such as methylation patterns that alter gene expression in response to stress, diet, or pathogen exposure; morphic fields ensure long-term coherence, preserving adaptive patterns across generations and populations even as epigenetic marks fade. The net result is a Lamarckian overlay on a Darwinian substrate: acquired characteristics influence the direction of evolution without altering the genes themselves, directing selection toward individuals already attuned to resonant solutions. Morphic resonance and natural selection constantly interact and influence each other, "in saecula saeculorum."

Additional examples illustrate the breadth of morphic resonance across diverse taxa. Japanese macaques on Koshima Island began washing sweet potatoes in 1953. Once 20 percent of a given troop had adopted the practice, it appeared in younger individuals overnight. Meerkats in the Kalahari developed a new trill to warn of novel predators. Distant clans reproduced the same alarm signal mere months later. Dolphins in Shark Bay began to use sponges as masks; unrelated pods of dolphins then re-invented the same method independently. These cases follow the same strict structure: local innovation, threshold breaching, and non-local propagation.

Philosophically, the implications of resonant morphic memory reach even wider, touching the core of how reality is understood. The mechanistic worldview, dominant since Descartes and Newton, portrayed nature as a vast machine, each organism isolated within its own skin, competing blindly for survival in a dead universe of matter. Morphic fields reveal a deeper connectedness that restores purpose and memory to the natural order. The past does not die; it remains present as an active influence, shaping the probabilities of the present through ongoing resonance. Memory is not confined to brains or DNA; it constitutes a property of nature itself, inherent in the organizing fields that give form to living systems. Three-dimensional time echoes Henri Bergson's durée créatrice: time is a creative manifold pregnant with possibility, shaped by the cumulative experience of the living rather than a mere container for entropic decay. Individual organisms participate in a larger evolutionary memory, where every innovation contributes to the species' developmental potential and enriches its collective inheritance.

Wolf packs coordinate hunts across vast territories, employing flanking maneuvers refined in one region and adopted in another without direct contact, allowing seamless adaptation to prey behaviour changes. Dolphin pods develop signature whistles that identify individuals and coordinate group activity; these acoustic dialects evolve collectively within the field, persisting beyond matrilineal lines and facilitating social cohesion in fluid ocean environments. Crows pass on tool-making traditions that persist for generations, with young birds mastering complex sequences—bending wires into hooks or using sticks to fish ants—more readily than their parents, demonstrating accelerated cultural transmission. Individual creativity feeds the species field, and the field in turn enables further creativity, creating a positive feedback loop of increasing sophistication. Evolution no longer appears as a blind struggle of isolated competitors but as a collaborative endeavour, guided by an immanent memory that builds upon itself.

The universe is not indifferent to the strivings of its creatures but attentive to their efforts. Every hard-won insight, every adaptive response, perseveres within the morphic field and offers to those who come after gentle guidance. A fledgling crow feels the subtle pull of ancestral practice in its first movements, as if the air itself carries the echo of past successes; an untrained rat senses the path already smoothed by countless trials before it, navigating a maze with an ease that belies its inexperience. The past leans close and whispers that this has been done before, offering reassurance and direction in moments of challenge. And the future, listening attentively to the resonant hum of history, rises and opens like the dawn, illuminated by the accumulated light of all that has come before in the grand tapestry of life.

Mathematical Annex: Inherited Learning in Animals

1. Behavioral Temporal Metric

$$ds^2 = g_{ij}\,\Delta b^i\,\Delta b^j$$

This metric quantifies functional proximity between behavioral phases in a three-dimensional space, enabling resonance only for homologous learning stages. It warps distances so conserved adaptations like alarm calls remain close across generations, minimizing drift in the morphic cloud.

2. Morphic Field Accumulation

$$M^i(t^k) = \int K^i{}_j(b^k, b'^k)\,\psi^j(b'^k)\,d^3b'$$

The morphic field aggregates stabilized patterns from all past learners, forming a collective memory accessible non-locally. Weighted by resonance, it enables offspring to inherit maze-solving or foraging skills without direct training or genetics.

3. Gaussian Weight

$$G(ds^2) = \exp\left(-\frac{ds^2}{2\sigma^2}\right)$$

This function creates fading influence, strongest for near-identical behavioral alignments and decaying smoothly with dissimilarity. It ensures rapid spread of adaptive behaviors like milk bottle opening while preventing resonance with unrelated patterns.

4. Resonance Kernel

$$K_{ij}(b^k, b'^k) = G(ds^2)\,\delta_{ij} + \kappa\,\partial_i\partial_j G(ds^2)$$

The kernel combines direct attraction to ancestral endpoints with curvature-guided paths along reinforced trajectories. Phase-specific resonance prevents cross-talk, canalizing populations toward adaptive norms like tool use in crows.

5. Geodesic in Morphic Cloud

$$\exp_p(v) = \gamma(1), \quad \frac{D\gamma}{d\lambda} = M^i(\gamma)\frac{\delta}{\delta b^i}$$

Populations evolve along the smoothest path defined by the morphic field, converging to principal adaptive trajectories. This geodesic guidance explains why untrained rats reconverge to efficient maze routes across distant labs.

6. Action Functional

$$S[\gamma] = \int \left[\frac{1}{2} g_{ij}\dot{\gamma}^i\dot{\gamma}^j - V_{\text{core}}(\gamma) + \kappa\,\delta\psi \cdot M(\gamma)\right] d\tau$$

Nature minimizes this action to favor smooth, historically aligned behavioral paths with low deviation from core adaptations. It drives inheritance by rewarding resonance with ancestral patterns, as seen in persistent tit behaviors.

7. Resonance Score

$$\mathcal{R} = \int \delta\psi_i(b^2, b^3)\, M^i(b^1, b^2, b^3)\, db^2\, db^3$$

This score evaluates trial behaviors against memory from matching early-phase ancestors, predicting inheritance potential. High resonance votes for efficient escapes in rats, suppressing maladaptive deviations across generations.

8. Equation of Motion

$$\frac{D^2\gamma^k}{d\tau^2} + \Gamma^k_{ij}\dot{\gamma}^i\dot{\gamma}^j = -\partial^k V_{\text{core}} + \kappa\, K^k{}_l\,\delta\psi^l$$

Inertia resists change while core adaptation and historical nudges pull populations toward resonant norms. This dynamics ensures robust repeatability of learned behaviors like predator avoidance in meerkats.

9. Morphic Field Update

$$\frac{\partial M^i}{\partial b^1} = \kappa\,\delta^{(3)}(b^k - \gamma^k(\tau))\,\psi^i(\gamma(\tau))$$

Successful adaptations stamp their final patterns into the field via Dirac delta, reinforcing early phases for future access. Only persistent behaviors like potato-washing in macaques contribute, enabling cumulative inheritance.

Mathematical Glossary

- ds^2: Behavioral distance squared; g_{ij}: Metric tensor warping space; Δb^i: Phase coordinate difference.
- M^i: Morphic field vector; K^i_j: Resonance kernel; ψ^j: Stabilized pattern descriptor; d^3b': Volume element over past paths.
- $G(ds^2)$: Gaussian weight; $\exp$: Exponential function; σ: Width parameter.
- δ_{ij}: Kronecker delta (1 if $i = j$, else 0); ∂_i: Partial derivative; κ: Guidance strength.
- γ: Evolutionary path; $\exp_p(v)$: Exponential map; $D/d\lambda$: Covariant derivative.
- $S[\gamma]$: Action functional; $\dot{\gamma}^i$: Velocity; V_{core}: Core potential; $\delta\psi$: Pattern deviation.
- $\mathcal{R}$: Resonance score; Γ^k_{ij}: Christoffel symbols for curvature.
- $\delta^{(3)}$: 3D Dirac delta for point updates; τ: Generational parameter.

IV

Morphic Resonance and the Emergence of Homo Sapiens

The origin of Homo sapiens is one of the most dramatic discontinuities in the entire primate fossil record. For nearly six million years after the last common ancestor, chimpanzees, hominins evolved slowly within the australopithecine and early Homo grades—bipedal, tool-using, but confined largely to Africa, with the exception of the Armenian highlands. Brain sizes rarely exceeding 600 cubic centimeters and behavioral repertoires limited to basic foraging, opportunistic scavenging, and rudimentary social grooming. Cranial vaults remained low and elongated, facial prognathism pronounced, brow ridges massive, and vocal anatomy constrained to a narrow range of guttural sounds.

Then, in a narrow window centered around 300,000 BC, a radically new form appeared across East, North, and Southern Africa: anatomically modern humans with high, globular braincases, delicate and retracted facial bones, prominent chins, reduced sexual dimorphism, and descended larynges enabling fully articulate speech. Within another 100,000 years, these beings dispersed globally, gradually replacing Neanderthals, Denisovans, Homo erectus, and relict Heidelbergensis populations with only trace genetic introgression in most regions. They arrived not as primitive migrants but as bearers of fully formed symbolic culture: pressure-flaked blade tools, composite bone harpoons, tailored clothing with eyed needles, long-distance exchange networks of obsidian, marine shells, and ostrich eggshell beads, and representational art—ochre engravings, carved ivory figurines, deliberate burials with red ochre and grave goods, and early musical instruments.

This transition defies the expectations of gradual Darwinian change. The anatomical shift was completed in less than 5 percent of hominin evolutionary history—less than 300,000 years out of over six million. The same suite of modern cranial, postcranial, and neural traits emerged nearly simultaneously in Ethiopia (Herto), Morocco (Jebel Irhoud), and South Africa (Florisbad, Border Cave) despite glacial barriers, arid corridors, and limited gene flow across the Sahara and Rift Valley. The transformation involved coordinated remodeling of hundreds of traits: cranial vault expansion to 1,350–1,500 cm^3, facial retraction and orthognathism, descent of the larynx and hyoid, narrowing and reshaping of the pelvic inlet, prolongation of childhood and delayed maturation, expansion of the prefrontal and temporal cortex, and the rewiring of Broca's and Wernicke's areas—all functionally integrated from their first appearance in the fossil record. The transition unfolded during repeated demographic bottlenecks, with effective population size plummeting below 10,000 individuals during Marine Isotope Stage 6 (195,000 to 123,000 years ago) and again during the Toba supereruption aftermath (74,000 years ago). Most strikingly, the genetic machinery required for modern cognition and anatomy was already present in Neanderthals, Denisovans, and even Homo heidelbergensis millions of years earlier, lying dormant until becoming abruptly and coherently expressed in Homo sapiens.

Conventional population genetics offers no satisfactory mechanism for this remarkable history. Synchronizing selection across thousands of independent loci would require correlation mechanisms that do not exist in the genome. Producing identical neural, skeletal, and vocal innovations in isolated African subpopulations demands a series of parallel mutations of vanishing probability—on the order of 10^{-15} per locus per generation. Population bottlenecks should erase adaptive variation through genetic drift and inbreeding depression, not sculpt coherent, multidimensional novelty. The presence of pre-existing but unexpressed developmental genes—conserved across archaics yet only deployed in Homo sapiens—points to a timing and coordination mechanism external to the genome itself, one that operates not through mutation and selection alone, but through resonance with past forms.

This essay proposes that the Homo sapiens transition follows the same six-phase process of morphic resonance within the three-dimensional temporal framework, as previously applied to the fin-to-limb transition, cichlid radiations, and Darwin's finches (see Chapter Two). The morphic field guiding hominin development—initially shaped by hundreds of millions of australopithecine, habiline, and heidelbergensis adults—directed embryos toward robust jaws, low cranial vaults, limited vocal range, and constrained behavioral plasticity. A new Homo sapiens field emerges through a minute initial perturbation in a founding population, gradual decay of ancestral influence over generations, self-reinforcement in small refugial populations under environmental stress, crossing of a reproductive isolation threshold, integration of novel traits through conserved developmental entanglement, and final partitioning of phenotypic space upon secondary contact with archaic hominins.

Phase 1: The Seed of Divergence

The process begins around 300,000 years ago in scattered African locales—Jebel Irhoud in the northwest, Herto and Omo Kibish in the east, and Klasies River Mouth in the south. Small founding populations inherit the parental morphic field of Homo heidelbergensis almost exactly but carry a tiny deviation—no more than eight-tenths of one percent—in the timing and duration of cranial vault expansion during late fetal and early postnatal development. This perturbation might arise from the founder effect as a handful of individuals are isolated by glacial drought cycles that fragment woodlands into refugial pockets, or from the rare survival of a developmental outlier under new ecological pressures—such as a child born during a period of abundant calorie-dense marine resources along the Cape coast, allowing prolonged brain growth beyond the ancestral norm. The deviation is small enough to preserve full Homo-grade functionality—robust mastication, effective thermoregulation, viable neonatal head passage through the birth canal—but large enough to establish a new direction in the multidimensional developmentspace of possible phenotypes.

These early populations are not yet recognizably modern. Skulls from Jebel Irhoud retain thick cranial bones, moderate supraorbital tori, and elongated vaults with occipital buns. Facial projection remains significant, and brain size hovers around 1,300 cm^3—within the upper heidelbergensis range but not yet fully globular. But the seed is planted: a fractional shift in the developmental trajectory that will, under the right demographic and ecological conditions, be amplified into a new stable attractor state.

Phase 2: The Fading Echo of Ancestry

For the first 100,000 years after the initial split, sapiens embryos continue to develop under the dominant guidance of the Heidelbergensis morphic field. The ancestral resonance—built from hundreds of millions of successful archaic adults across Africa and Eurasia—overwhelms the faint signal from the few deviant founders. This influence decays in two complementary ways. First, as the developmental pathways of parent and daughter lineages begin to differ—braincases rounding, faces flattening, larynges descending—the similarity between their embryonic trajectories decreases, producing a smooth, Gaussian fall-off in influence. Second, with each passing generation, the simple passage of time exponentially reduces the ancestral echo, as the temporal distance in the first time dimension (generational succession) grows.

During this period, hybrids between early sapiens and archaic populations remain viable, often displaying mosaic characteristics: high vaults combined with large brow ridges, modern chins with robust mandibular corpora, or descended larynges in otherwise archaic vocal tracts. The parental influence remains strong enough—above roughly forty percent of its initial value—that embryonic development receives coherent, if blended, instructions. Fossil evidence from Skhul and Qafzeh (120,000 to 90,000 years ago) supports this: early dispersers into the Levant show mixed traits, interbreeding with local Neanderthals. Only as the kernel of ancestral influence weakens below ten percent, around 200,000 years ago, do purely sapiens morphologies begin to dominate in African sites like Omo Kibish and Herto. This gradual uncoupling explains the long overlap phase seen in the fossil record, with transitional forms like the Florisbad skull (260,000 years ago) showing intermediate vault height and facial angle.

Phase 3: Building the New Consensus

Between 300,000 and 70,000 years ago, the daughter sapiens field begins to shape its own future. With each generation, the field incorporates the adult forms of individuals that survive and reproduce. These contributions are weighted by how closely each individual's embryonic journey matched that of the current cohort—ensuring that only relevant past successes influence present development. In small, bottlenecked populations—coastal South Africa (Blombos, Pinnacle Point), Rift Valley lakes (Lake Turkana), Ethiopian highlands (Dire Dawa), and North African oases—every successful adult exerts outsized influence. A single individual at Blombos Cave 77,000 years ago who regularly used red ochre in symbolic patterns, crafted bifacial stillbay points, and wore perforated shell beads contributes disproportionately to the emerging field.

Environmental stress simultaneously increases developmental noise. The supereruption of Toba (74,000 years ago), Heinrich events, and prolonged megadroughts elevate variation in cell division timing, hormone pulses, and growth factor gradients. This noise generates rare variants: slightly more descended larynges enabling expanded phonetic range, enhanced hand precision via modified intrinsic muscle insertions, denser theory-of-mind neural circuitry in the temporoparietal junction, and prolonged sensitive periods for language acquisition. When these variants prove adaptive—enabling better cooperative hunting, pedagogical transmission, or abstract planning—they are rapidly reinforced. Within a few dozen generations, configurations like high-vaulted skulls with fully modern brow ridges, chins, and vertical foreheads begin to dominate the daughter field, gradually overwhelming the fading Heidelbergensis signal.

The process is self-accelerating. The more individuals achieve the modern form, the more strongly the field guides subsequent embryos toward that same outcome. Populations in separate refugia converge on nearly identical cranial, postcranial, and behavioral endpoints despite different starting variants, because the reinforcement process draws them toward the same viable configurations preserved in their shared ancestral resonance. The Stillbay and Howiesons Poort technocomplexes appear synchronously across southern and eastern Africa, despite isolation, reflecting this field-level convergence.

Phase 4: Crossing the Threshold of Isolation

By 195,000 years ago, during the severe bottleneck of Marine Isotope Stage 6, the normalized overlap between the sapiens and Heidelbergensis morphic fields falls below a critical value—approximately seventy percent. The overlap measures how consistently the guidance vectors of the two fields point in the same direction across all stages of development, normalized by the internal consistency of each field. When overlap remains high, hybrid embryos receive coherent, blended instructions and develop viable mosaic forms. Below the threshold, the instructions conflict fundamentally: one field directs full laryngeal descent for complex speech, the other retains ape-like vocal constraints; one specifies 1,500 cm^3 vaults with thin tables, the other caps growth at 1,250 cm^3 with thick bones; one pulls for prolonged childhood and delayed dental eruption, the other enforces rapid maturation.

The resulting developmental chaos produces non-viable or infertile hybrids. Homo sapiens achieves reproductive isolation despite later geographic overlap with Neanderthals and Denisovans. Genomic studies confirm this: while limited interbreeding occurred after 70,000 years ago, fertile offspring were relatively rare and backcrossed primarily into sapiens populations. The speciation threshold marks the point of no return—the moment two lineages become distinct species in both form and developmental destiny.

Phase 5: The Birth of Integrated Novelty

As the speciation progress parameter rises from zero to one between 300,000 and 50,000 years ago, the modern adult phenotype emerges not as a patchwork of isolated traits but as a functionally coherent suite. The adult form of a Homo sapiens individual is constructed from three components. The first is the ancestral Heidelbergensis configuration—robust skeleton, elongated vault, prognathic face—weighted by the square root of one minus the progress parameter. The second is a pure novelty component—such as globular vault shape, chin prominence, or descended hyoid—weighted by the square root of the progress parameter. The third is a cross-term that reaches maximum strength halfway through the transition (around 150,000 years ago) and couples correlated structures through a matrix derived from deeply conserved developmental relationships.

At this midpoint, the entanglement term peaks. A globular vault does not appear in isolation; it emerges alongside thinned cranial bones to reduce weight, expanded parietal bosses for larger temporal lobes, and a more vertical forehead for enhanced prefrontal volume. A chin forms concurrently with facial retraction, vertical symphysis, and reduced mandibular robusticity—all linked through shared growth fields in the mandibular and maxillary regions. Laryngeal descent couples with expansion of Broca's and Wernicke's areas, hyoid bone reshaping, modified thoracic innervation for breath control in speech, and enlarged thoracic vertebrae for upright posture support.

This developmental entanglement, rooted in ancient pathways like neural crest migration, Hox gene expression, BMP and FGF signaling, and endothelin pathways, guarantees that innovations are viable from their first appearance. No trait evolves alone; the system ensures functional harmony. The result is the sudden appearance of the full modern anatomical syndrome in fossils like Omo I (195,000 years ago) and Skhul V (100,000 years ago)—complete, integrated, and immediately adaptive for symbolic communication, pedagogical culture, and ecological dominance.

Phase 6: Partitioning in Sympatry

Between 100,000 and 70,000 years ago, sapiens populations disperse out of Africa via southern and northern routes and enter secondary contact—sympatry—with Neanderthals in the Levant and Europe, Denisovans in Central and Southeast Asia, and relict Heidelbergensis or Homo erectus populations in Indonesia and China. In these zones of overlap, developmental trajectories actively curve toward regions of phenotypic space that are dense with successful Homo sapiens forms and away from areas still dominated by archaic resonance. The strength of this repulsion is proportional to the gradient of remaining field overlap.

In the Levant 100,000 years ago, early sapiens exaggerate symbolic material culture—ochre use, shell beads, incised plaques—while Neanderthals retain static Mousterian tools. In Southeast Asia, sapiens develop edge-ground axes, seafaring canoes, and tropical forest adaptations while Homo floresiensis persists in isolation on Flores. Upon secondary contact in Europe 45,000 years ago, Homo sapiens accelerate artistic and technological output: Chauvet Cave paintings with perspective and motion, ivory Venus figurines, bone flutes with tuned scales, and tailored clothing with bone needles appear within a few millennia.

The Homo sapiens morphic cloud separates into a distinct, expanding basin of phenotypic possibility. The archaic cloud contracts and collapses within 15,000 years. Neanderthals, despite robust physiques and fire mastery, cannot access the Homo sapiens developmental attractor—their embryos remain locked into the ancestral field. The partition is final: sapiens dominate, while archaics gradually vanish.

The Deeper Pattern

The Homo sapiens transition is not a random walk through genetic space but a resonance-guided recognition of a latent possibility within the hominin manifold. Somewhere in the vast temporal landscape—perhaps in a Heidelbergensis child who first drew geometric patterns in ochre 500,000 years ago at Twin Rivers, or spoke in recursive syntax 400,000 years before at Gesher Benot Ya'aqov—the modern configuration was fleetingly achieved. Hundreds of millennia later, when bottlenecked populations, developmental stress, and ecological opportunity aligned, the accumulated resonance activated this dormant form.

The process explains the otherwise puzzling patterns of human evolution. The rapidity of the transition arises from threshold dynamics: tiny initial perturbations remain latent until enough successful individuals accumulate to push the system past the critical overlap value, after which change accelerates exponentially. Geographic synchrony across Africa occurs because all daughter populations draw from the same ancestral resonance cloud; local variants are selected, but self-reinforcement converges on the same viable endpoint. Population bottlenecks facilitate innovation by increasing the influence of each successful adult and elevating developmental noise through stress—transforming demographic crises into opportunities for rapid field restructuring. The long delay between the evolution of cognitive genes and their phenotypic expression reflects the requirement for sufficient resonance from actually realized forms before the field can activate dormant genetic potential.

In this view, Homo sapiens is less an invention ex nihilism and more the result of innovation empowered by morphic memory. The globular skull, the speaking tongue, the imagining mind—these were not built from scratch but assembled by drawing on the archive of past forms. The six-phase process traces this emergence with precision. Humanity in its modern form was always present as a potential in the morphic cloud, waiting for the right moment to emerge.

Mathematical Annex: The Six Equations (Adapted for *Homo sapiens*)

1. Daughter Field Initialization

$$M_S^i(t_{\text{split}}^k) = M_H^i(t_{\text{split}}^k) + \epsilon \, \Delta M_{\text{cranium}}^i(t_{\text{split}}^k)$$

The sapiens morphic field begins as a near-copy of the *Homo heidelbergensis* field with a small perturbation ($\varepsilon \approx 0.008$) in cranial vault expansion timing, introduced by founder effects or rare developmental variants in isolated African populations around 300,000 years ago.

2. Divergence Kernel

$$K_{S \leftarrow H j}^i = \exp\left(-\frac{ds_{SH}^2}{2\sigma_{\text{div}}^2}\right) \delta_j^i \, e^{-\lambda(t^1 - t_{\text{split}})}$$

The fading influence of the ancestral *heidelbergensis* field is quantified as sapiens cranial, facial, and neural trajectories diverge over 100,000 years, decaying exponentially with generational time and reduced developmental similarity, dropping below 10% by 200,000 years ago.

3. Daughter Field Evolution

$$\frac{\partial M_S^i}{\partial t^1} = \kappa_S K_{Sj}^i \phi_S^j + \kappa_{S \leftarrow H} K_{S \leftarrow H j}^i \phi_H^j + \eta \xi^i(t^k)$$

The sapiens field strengthens with each generation as successful adults in small refugial populations (e.g., Blombos, Rift Valley) contribute disproportionately, amplified by stress-induced developmental noise that generates viable variants in speech, cognition, and dexterity.

4. Speciation Threshold

$$\mathcal{O}_{SH} = \frac{\int M_S^i M_H^i \, d^3t}{\sqrt{\int M_S^j M_S^j \, d^3t \cdot \int M_H^k M_H^k \, d^3t}} < \theta_c$$

Reproductive isolation is achieved when normalized overlap between sapiens and archaic morphic fields falls below ~70% around 195,000 years ago, causing conflicting developmental instructions in hybrids and rendering interbreeding non-viable.

5. Form Bifurcation

$$\phi^i_S = \sqrt{1-p}\,\phi^i_H + \sqrt{p}\,\phi^i_{\text{modern}} + \sqrt{p(1-p)}\,\Phi^{ij}_{\text{entangle}}\phi^j_H$$

The modern sapiens phenotype emerges as a coherent blend of ancestral structure, novel traits (e.g., chin, globular vault), and developmentally coupled components (e.g., larynx + Broca's area), with maximum integration at the midpoint (150,000 years ago).

6. Morphic Cloud Partition

$$\frac{D^2\gamma^k_S}{d\lambda^2} = \kappa_S K^k_{Sl}\delta\phi^l_S - \kappa_{S\leftarrow H}\nabla^k \mathcal{O}_{SH}$$

In sympatry after 70,000 years ago, sapiens developmental trajectories accelerate away from archaic hominins (Neanderthals, Denisovans), driven by attraction to successful sapiens forms and repulsion from residual ancestral field overlap, culminating in rapid cultural and anatomical divergence.

V

Morphic Resonance and the Development of Early Human Cognition

Human consciousness did not arrive in a single thunderclap of genius. It emerged gradually, like dawn light spreading across a vast savanna, first as a faint glow on the horizon, then as a steady illumination that revealed the world in ever-finer detail. For millions of years, our ancestors lived inside a shared mental space—a collective cloud of memory and intuition that bound individuals together without words. This was the parental field of cognition, a shimmering web of telepathic resonance where thoughts, emotions, and skills flowed freely between minds. A mother's discovery of a new way to crack open a bone for marrow did not remain hers alone; it rippled outward, gently nudging others toward the same insight, even if they stood hundreds of miles away. No one taught, no one learned in the modern sense. The knowledge was simply held in common, accessible to all who tuned their attention to the hum of the group.

Gradually a new type of cognition emerged, focused on individual rather than collective consciousness. This essay traces this long journey in six precise stages, each corresponding to a transformation within the morphic cloud of humanity. We begin with the first faint perturbation—a tiny innovation that disturbed the smooth surface of collective awareness. We then trace how ancestral influence slowly faded, how the new pattern gathered strength through repetition and environmental accident, how it finally broke free into a separate cognitive lineage, how it blended old and new into a hybrid mind, and how, at last, the cloud itself divided into two realms: one of articulate speech and private thought, the other a quieter, deeper current of shared feeling and cultural memory. Stone tools and language did not evolve in parallel by chance. They were twin expressions of the same resonant process, each shaping the other as cognition flowed from cloud to individual minds.

Stage One: The First Whisper – A Tiny Innovation in the Primal Cloud

Picture a dry riverbed in what is now Ethiopia, three million years ago. The sun beats down on a small band of Australopithecus or early life Homo habilis hominids. One individual—let us imagine a young female—picks up a fist-sized lump of volcanic rock. She has done this before. Her companions have done it thousands of times. The motion is automatic: strike the core with a hammerstone, detach a flake, use the sharp edge to slice meat or scrape hide. The action is efficient, but not elegant. The flakes are irregular, the edges uneven. Most strikes waste material. Most flakes are too thick to be useful.

On this particular afternoon, something shifts. Perhaps her grip slips. Perhaps the angle of the hammerstone catches the light differently. Whatever the cause, the blow lands at a slightly steeper angle than usual. A long, thin flake peels away—clean, sharp, almost delicate. She turns it in her hand, surprised by its lightness. The edge slices through tendon like a modern knife. In that moment, a new possibility is born.

This is the first stage: the initialization of a daughter field. The innovation is minuscule—a change in wrist angle, a fraction of a degree in platform preparation. In a purely individual brain, it might have been forgotten by nightfall. But early hominins did not think alone. They thought inside the morphic cloud, a field of shared resonance that stored patterns of behavior the way a quantum field stores information. The new angle registered not just in her motor cortex but in the collective memory of morphic resonance. Within days, others in her band began producing similar flakes. Within months, the pattern appeared at distant sites—Gona, Hadar, Lokalalei—places separated by rivers, mountains, and hundreds of kilometers of uninhabited bush. No one walked between them. No one carried a sample flake as a model. The knowledge traveled through the morphic cloud, carried by resonance.

Archaeologists call these artifacts Oldowan, after the gorge where they were first studied. To the untrained eye, they look crude: broken cobbles, jagged edges, no symmetry. But beneath the roughness lies a quiet revolution. The best Oldowan flakes are remarkably consistent in length, thickness, and cutting angle. That consistency cannot be explained by chance or by direct teaching. It points to a distributed intelligence—a mind larger than any single cerebral cortex or skull.

Stage Two: The Fading Echo – Ancestral Instinct Loses Its Grip

For half a million years, the Oldowan pattern held sway. Bands across East Africa and as far away as the Armenian highlands produced the same basic tools using the same basic techniques. The morphic cloud reinforced continuity. A child learned by watching, yes—but also by feeling the rightness of a motion, the way a violinist feels the correct bow pressure without being told. The parental field—built from millions of prior strikes—exerted a gentle but persistent pull toward the familiar.

Then, around 1.7 million years ago, the pull began to weaken. Homo erectus appeared on the landscape—taller, longer-limbed, with a larger brain and a restless curiosity. They did not abandon the old ways entirely. They still made choppers and flakes. But now, some individuals began shaping stones on both sides, removing flakes from opposite faces to create a rough symmetry. The resulting tools—handaxes—were heavier, more versatile, and far more difficult to make. Each blow had to be planned. The final shape had to be held in mind before the first flake fell.

This is the second stage: divergence. The ancestral echo—the Oldowan way—did not vanish overnight. It lingered like a memory of childhood. A knapper might begin with the familiar single-faced core, then hesitate, then flip the stone and remove a flake from the other side. The new pattern competed with the old. Sometimes it failed. Sometimes the stone shattered. But when it succeeded, the resonance was powerful. A well-made handaxe could butcher an elephant, dig roots, or serve as a weapon. Its utility amplified the signal.

The divergence spread slowly at first. At sites like Konso in southern Ethiopia, handaxes appear alongside Oldowan cores—a palimpsest of overlapping traditions. Over time, the handaxe pattern grew stronger. By 1 million years ago, it dominated in Europe, Asia, and Africa. At Boxgrove in England, 500,000 years ago, knappers left behind quartzite handaxes so symmetrical they look almost modern. At Olorgesailie in Kenya, the same forms appear in the same time frame. The parental field still whispered—"strike here, like this"—but its voice grew fainter with each generation. The daughter field was learning to speak for itself.

Stage Three: Self-Reinforcement and the Spark of Accident

By 700,000 years ago, the daughter field had a life of its own. Handaxes were no longer rare experiments. They were the norm. Knappers now shaped stones with confidence, removing flakes in predetermined sequences to achieve a desired edge angle or weight distribution. The process was still silent—no words, no diagrams—but it was no longer purely intuitive. It required foresight. It required memory. It required a mind that could hold multiple steps in superposition.

This is the third stage: evolution through self-reinforcement and stochastic innovation. Every successful handaxe strengthened the pattern. Every apprentice who watched a master knapper absorbed not just the motion but the "feel" of correctness. The resonance was now local—within a band, within a valley—but it was intense. A single skilled individual could shift the entire group's technique in a season.

Accident played its part. A lightning strike might fracture a boulder into perfect blanks. A flood might expose a vein of fine-grained flint. A child playing with stones by a fire might discover that heat altered their fracture properties. These events may have been random, but they were not wasted. The morphic cloud absorbed them, tested them, and—if they worked—amplified them. At Zhoukoudian in China, 700,000 years ago, Homo erectus used fire to heat-treat silcrete, producing tools with glass-like edges. At Gesher Benot Ya'aqov in Israel, 780,000 years ago, the same technique appears. No migration can explain the synchrony. The cloud carried the spark of innovation across vast distances.

The daughter field was now a storm gathering force. It fed on success, on repetition, on the quiet accumulation of small improvements. And it began to demand more than the old morphic telepathic medium could provide.

Stage Four: The Breaking Point – When Cloud and Cortex Part Ways

Around 300,000 years ago, the storm broke. In caves and rock shelters from Morocco to the Levant, a new kind of tool appeared: the Levallois core. Unlike the handaxe, which was shaped by removing flakes until the core itself became the tool, the Levallois technique treated the core as a factory. The knapper first prepared a tortoise-shaped platform, then struck a single predetermined flake that carried the desired shape—long, thin, symmetrical.

The process required nine distinct steps, each dependent on the last. It required abstraction. It required language, or something very close to it.

This is the fourth stage: speciation. The daughter field had grown so complex that it could no longer be contained within the parental cloud. The old telepathic resonance—perfect for sharing hunger, fear, or the location of a fruit tree—faltered under the weight of sequential logic. You cannot "feel" the correct angle for a Levallois platform. You cannot intuit the convexity needed for a laminar flake. You must explain it instead in logical, rational terms and then demonstrate the proper technique for producing it. Moreover, the key parts must be named: core, crest, platform, and bulb of percussion. Thus, rational analysis and language are necessary.

The human anatomical hardware was ready by this time. The larynx had descended. Broca's area had expanded. But hardware is not enough. The software—the cognitive architecture—had to cross a threshold. And so it did. At sites like Blombos Cave in South Africa, 100,000 years ago, we find ochre engraved with geometric grids. At Qafzeh in Israel, 92,000 years ago, we find perforated shells strung as beads. These are not tools, but rather symbols. They are evidence that the daughter field had achieved the cognitive equivalent of reproductive isolation and could propagate itself independently.

This cognitive speciation was not cruel. It did not erase the past. It simply declared independence. The parental field retreated to the margins—the dreams of shamans, religious rituals, the unspoken bonds of clan kinship. The daughter field claimed the center, deploying its core skills of planning, teaching, and storytelling. The modern human mind was born.

Stage Five: The Hybrid Mind – Ancestral Intuition Meets Linguistic Precision

The Upper Paleolithic, beginning around 50,000 years ago, is the fifth stage: bifurcation into a hybrid form. The daughter field did not reject its ancestry. It embraced it instead. The result was a mind that could do two things at once: think in rational sequences and feel in morphic synchrony.

And so at Lascaux, 17,000 years ago, artists painted horses in mid-gallop, using charcoal strokes to suggest motion. The technique required linguistic planning—"first the outline, then the shading, then the highlight"—but the emotional impact came from ancestral morphic empathy. The painter "felt" the horse's power, its fear, its grace. At Dolní Věstonice in Moravia, 26,000 years ago, ceramic figurines were fired in kilns that required precise temperature control. The technology was daughter-field logic. The choice of subject—fertility, motherhood, the cycle of life—was parental-field resonance drawn from Morphic resonance.

Language was the bridge. It allowed the daughter field to name the parts of a tool, the steps of a hunt, the rules of a ritual. But it also allowed the parental field to speak through metaphor, myth, and song. A hunter could plan a drive in detail—"we will station archers here, beaters there"—and still enter a trance state to "become" the prey. A craftsman could calculate the tensile strength of sinew thread and the weight of beads or amulets still believe the necklace was an amulet inhabited by a spirit.

This hybridity explains the cultural explosion of the Upper Paleolithic. Bone flutes appear in Germany. Venus figurines appear in Russia. Atlatls—spear-throwers—appear in France and Peru within a few thousand years of each other. The daughter field provided the precision. The parental field provided the synchrony. Together, they produced behavioral modernity.

Stage Six: The Great Partition – Speech and Silence Divide the Cloud

The final stage is partition. By 10,000 years ago, the daughter field had won. Agriculture, villages, writing, mathematics—all were its children. The parental field did not disappear, it simply moved underground not the world of the unconscious. It became the realm of dreams, of art, of the unspoken understanding between friends and family members, between spouses and parents and their children. It became the morphic hum beneath the surface, constantly influencing culture on the instinctive level.

The partition was active, not passive. The daughter field pulled away, repelled by the imprecision of pure resonance. It demanded clarity, repeatability, and falsifiability. The parental field pushed back, preserving cohesion, empathy, and the sense of belonging to something larger than self. The tension between these two modes of cognitions lingers and still informs the human condition today.

We experience this cognitive tension every day. A scientist writes a paper with equations and citations—pure daughter-field logic. A poet writes a sonnet that makes strangers weep—pure parental-field resonance. A programmer debugs code while listening to music that synchronizes his heartbeat with a thousand others. The morphic cloud is still there, uniting us all, albeit thinner and more diffuse.

Biological and Archaeological Corroborations

From a biological standpoint, the timeline of cognitive advancement poses a profound puzzle. Genetic evidence suggests that the mutations required for modern symbolic behavior—such as the neurological wiring for recursive language or abstract planning—should take hundreds of thousands of years to spread through a population. Yet the archaeological record shows a dramatic acceleration: simple ochre markings at 100,000 years ago give way to intricate engravings, beadwork, and representational art by 70,000–40,000 years ago. This leap occurs far too quickly to be explained by standard evolutionary mechanisms alone.

The six-stage resonant model offers a solution. Rather than relying solely on slow genetic drift, cognitive innovations spread through the morphic cloud at the speed of recognition. A rare neurological variant enabling better mental rotation, for example, does not need to become common through reproduction. Once one individual demonstrates its utility—say, by consistently producing superior handaxes—the pattern resonates outward, making similar neural configurations more probable in others, even without direct genetic inheritance. Over generations, selection pressures favor brains that can *tune in* to these resonant fields, effectively lowering the threshold for complex cognition. Cranial endocasts from early *Homo sapiens* show gradual enlargement of Broca's and Wernicke's areas, but the model predicts that functional demand precedes structural change: the need to articulate and teach sequential processes reshapes the brain, not the other way around.

Archaeologically, the evidence for non-local transmission is equally compelling. Acheulean handaxes, which dominate the record from 1.7 million to 200,000 years ago, exhibit remarkable consistency across vast distances. Tools from Swanscombe in England, Isimila in Tanzania, and the Kapthurin Formation in Kenya vary by less than five percent in length-to-width ratio, edge angle, and bilateral symmetry—despite being separated by oceans and millennia. Traditional explanations invoke long-distance migration or trade, but no intermediate forms or transport routes support this. Instead, the uniformity suggests a shared cognitive template sustained by resonance.

The pattern repeats with later technologies. The Levallois prepared-core technique emerges almost simultaneously around 300,000 years ago in North Africa, the Levant, and Western Europe. Within 50,000 years, it appears as far east as India and as far south as the Cape. Bladelet production, heat treatment of silcrete, and microlithic composites follow the same synchronized trajectory, appearing in Siberia, South Africa, and Australia within narrow temporal windows. Migration models collapse under the weight of these coincidences; the morphic cloud, operating through the six stages, provides a coherent alternative. Resonance does not require physical contact—only similarity of intent, environment, and resonant potential.

Taken together, biology and archaeology paint a picture not of isolated genius but of distributed emergence. The modern mind did not evolve in one place and spread. It crystallized across the species, guided by an invisible field that amplified the possible into the actual.

Philosophical and Cultural Reflections

The six-stage journey from collective cloud to partitioned mind reframes what it means to be human in profound ways. We are accustomed to thinking of consciousness as private, bounded by the skull, a solitary flame flickering in the dark. The resonant model suggests otherwise: we are nodes in a field, individual yet inseparable from the whole. Our thoughts do not begin and end with us. They ripple outward, shaping distant minds as surely as gravity bends light. The daughter field—our capacity for logic, language, and deliberate planning—feels like freedom. The parental field—our inheritance of empathy, ritual, and unspoken knowing—feels like belonging. Both are essential. Neither is complete without the other.

This duality echoes ancient wisdom. Mystical traditions speak of a universal mind, a web of consciousness that binds all beings. Indigenous cultures describe knowledge arriving in dreams or visions, untaught yet true. Modern science, with its emphasis on measurable mechanisms, has often dismissed these accounts as metaphor. The resonant model bridges the gap: it proposes a physical process—non-local field interaction—that could underlie what mystics have long intuited.

Culturally, the implications are vast. The internet, for all its algorithms and servers, behaves like a new morphic cloud. A dance move invented in Seoul spreads to Lagos in hours. A scientific paradigm shift in one lab influences experiments on another continent before the paper is published. Memes, trends, and moral panics propagate with the speed and unpredictability of resonance, not diffusion. We call it "going viral," but the mechanism may be older than language itself.

The partition between daughter and parental fields carries moral weight. If thoughts shape distant realities, then intention matters. Compassion practiced in solitude reverberates globally, strengthening the parental field's capacity for cohesion. Hatred does the same, amplifying division. The equations are not abstract formalism—they are a physics of moral responsibility. Every act of teaching, every story told, every tool crafted is a vote for the kind of field we wish to inhabit, full of grace and wisdom.

In the end, the six stages are not a linear progression from primitive to advanced. They are a spiral, turning ever deeper into the mystery of mind. We speak now with individual voices, but we still dream in chorus.

A Resonant Legacy in Six Movements

From that first unexpected stone flake, sharp enough to cut tendons and resonant enough to change a species, to the glowing digital rectangles we now all hold in our hands, human cognition has unfolded as a single, unbroken symphony in six movements. The morphic cloud was our cradle, our tutor, our silent companion throughout this process. It carried us from raw instinct to luminous imagination, from the wordless unity of telepathic communion to the bright, articulate solitude of individual minds. Stone tools and spoken sentences were born as twins and nurtured by the same resonant morphic cloud, growing stronger with every shared discovery.

We speak now in private voices and plan in careful sequences. We teach with words that travel farther than any footstep ever could. Yet beneath the bright surface of daily thought, the old morphic currents still run—gentle, persistent, alive. They surface in laughter that rises unbidden in a crowded room. They arrive in the grief that shakes us before tragic news arrives. They manifest as a sudden intuition that solves a problem we did not yet know we faced. The cloud has grown quieter, yes, but it has never dissipated or dissolved.

We are the daughter field, shaped by logic and language, proud masters of the articulate present. Thanks to morphic resonance, we are also the living echo of every mind that came before us—every hand that struck a core, every eye that watched the stars and wondered. The six stages of cognitive emergence are not a ladder that we have climbed and abandoned. They are a spiral, turning ever outward, weaving the past into the future with every resonant heartbeat.

Mathematical Annex: The Emergence of Human Cognition in Six Stages

1. **Daughter Field Initialization**

$$M_D^i(t_{\text{split}}^k) = M_P^i(t_{\text{split}}^k) + \epsilon \, \Delta M_{\text{mutation}}^i(t_{\text{split}}^k)$$

At the instant a novel insight appears, the new (daughter) morphic field is born as an almost perfect copy of the ancestral (parental) field. A tiny perturbation—barely larger than a single improved wrist motion—is all that is needed to seed an entirely new cognitive lineage.

2. **Divergence Kernel**

$$K_{D \leftarrow Pj}^i = \exp\left(-\frac{ds_{PD}^2}{2\sigma_{\text{div}}^2}\right) \delta_j^i \, e^{-\lambda(t^1 - t_{\text{split}})}$$

This equation measures how strongly the ancestral way of thinking can still tug on the emerging way. Its value drops rapidly as the two patterns grow different and as ordinary generations pass, gently releasing the past's grip on the future.

3. **Daughter Field Evolution**

$$\frac{\partial M_D^i}{\partial t^1} = \kappa_D K_{Dj}^i \phi_D^j + \kappa_{D \leftarrow P} K_{D \leftarrow Pj}^i \phi_P^j + \eta \xi^i(t^k)$$

The new cognitive pattern strengthens through its own successful repetitions and whatever faint ancestral echoes remain. Sudden environmental shocks or lucky accidents add unpredictable sparks that can dramatically accelerate its growth.

4. **Speciation Threshold**

$$\mathcal{O}_{DP} = \frac{\int M_D^i M_P^i \, d^3t}{\sqrt{\int M_D^j M_D^j \, d^3t \cdot \int M_P^k M_P^k \, d^3t}} < \theta_c$$

When the similarity between the old collective mind and the new linguistic mind falls below a critical level, the two can no longer fully interbreed cognitively. This marks the irreversible birth of individual, speech-based thought.

5. **Form Bifurcation**

$$\phi^i_D = \sqrt{1-p}\phi^i_P + \sqrt{p}\phi^i_{\text{novel}} + \sqrt{p(1-p)}\Phi^{ij}_{\text{entangle}}\phi^j_P$$

The mature Upper Paleolithic mind is a living superposition: mostly new linguistic and symbolic capacities, yet still rooted in ancestral intuition. A subtle entanglement term ensures that innovations in one corner of the world instantly correlate with distant populations.

6. **Morphic Cloud Partition**

$$\frac{D^2\gamma^k_D}{d\lambda^2} = \kappa_D K^k_{Dl}\delta\phi^l_D - \kappa_{D\leftarrow P}\nabla^k\mathcal{O}_{DP}$$

The final equation describes the active splitting of the once-unified morphic cloud into two distinct realms. One realm becomes the bright, articulate world of individual speech; the other retreats into the quieter, ever-present domain of collective memory and dream.

VI

Resonant Geodesics: the Evolution of Stone Tools in Three Dimensional Time

In the shadowed dawn of human ingenuity, when the first sparks of consciousness flickered against the vast cognitive darkness of prehistory, stone tools emerged as the first steady lights in the unfolding story of the awakening of the human mind. These humble artifacts, chipped from flint and basalt, obsidian and quartzite, were not mere instruments of survival but vessels of a deeper communion—a resonance that echoed across epochs, binding distant knappers in an invisible web of shared possibilities. Long before the written word or the forge's glow, the evolution of these tools traced a path through three-dimensional time, where ordinary chronology intertwined with realms of potential designs and transitions. In this manifold, the shape of a handaxe's curve in Kenya might whisper to a counterpart in Spain, separated by fifty millennia yet drawn inexorably close by the affinity of form and technique. Such is the poetry of morphic resonance, a field that accumulates the triumphs of countless artisans, weighting each success with a gentle persistence that defies the linear march of time.

Imagine the Oldowan horizon, some 2.6 million years ago, when the earliest hominins—perhaps Australopithecus garhi or the dawn of Homo habilis—grasped river cobbles and struck them with instinctive force. These pebble tools, crude and opportunistic, marked the genesis of intentional craftsmanship. A core might yield a single sharp flake, its edge glinting like a promise under the African sun. Yet even in this primordial simplicity, a subtle metric governed the closeness of events. Two knappers, one in the rift valleys of Ethiopia and another amid the savannas of Tanzania, might produce near-identical choppers despite vast spatial gulfs, for their actions converged in the development spaces of possibility and ease of transition. The stretch of time along the chronological axis paled against the compression in design and method; what seemed eons apart in calendar time collapsed into intimate proximity, fostering a resonance that amplified the likelihood of repetition. Thus, the Oldowan tradition persisted not through rigid instruction but via an accumulating cloud of morphic memory, where each successful strike contributed a vector of form—length, width, edge acuity—integrated across the manifold's volume.

As millennia unfolded, this morphic cloud grew denser, its influence manifesting in the Acheulean symphony that blossomed around 1.7 million years ago. The handaxe, that iconic stone sculpture of grace and utility, bifacially worked to symmetrical perfection, represented a quantum leap in conceptual thinking. Homo erectus, wandering from Africa to Eurasia, carried within their neural pathways not just genetic inheritance but an attunement to this morphic field. Consider a knapper in the Olorgesailie basin, selecting a lava cobble and envisioning the bilateral removal of flakes to achieve equilibrium. The field, built from integrals over past successes, weighted contributions through a Gaussian veil that decayed gracefully with distance in the manifold. A handaxe crafted 400,000 years prior might still exert a fifth of its original pull, nudging the present artisan toward refined symmetry. This was no mere imitation; a kernel of morphic resonance operated dually, directly reinforcing exact matches while subtly guiding toward enhancement via a secondary term that created attractors for improvement. The knapper's hands, callused and precise, felt an almost gravitational tug, pulling asymmetrical attempts into the well of graceful bilateral forms.

In this three-dimensional temporal landscape, stable traditions etched geodesics—the shortest paths across the morphic resonance manifold, propelled by the field's directional force. The Levallois technique, emerging around 300,000 years ago in the Middle Paleolithic, exemplifies such a groove. Originating in the Levant, East Africa, or perhaps the Armenian highlands, it involved meticulous preparation of a core to determine flake detachment, yielding points and blades of predetermined shape. Starting from an initial vector of preferential flaking, the morphic push maintained parallel velocity along the curve, sustaining the method across vast stretches of ordinary time—300,000 years of continuity from the deserts of Syria to the caves of France and the rock shelters of Armenia. Knappers in disparate regions, unaffected by direct contact, converged on laminar reduction sequences, their cores turtle-backed and faceted, as if drawn along by an invisible vector. The action along potential paths minimized cost: rapid shifts incurred kinetic penalties, impossible forms clashed against cognitive potentials, but alignment with the resonant cloud granted profound discounts, making the Levallois school the path of least resistance.

Yet evolution demands divergence, and the morphic framework accommodates speciation with elegant subtlety. When a tradition branched—say, the Mousterian variants among the Neanderthals parting from African Middle Stone Age lineages—a daughter field initialized as a near-copy of the parent, infused with a minute perturbation, perhaps five percent in amplitude. This twist, random yet bounded, set the stage for independent growth. The parent's echo lingered, decaying exponentially post-split, while the daughter reinforced itself through local successes and a veil of noise that introduced serendipitous variation. Over time, overlap diminished; a cosine measure of field similarity dipped below a critical threshold, heralding a new archaeological culture. Thus, the Prepared Core Technologies of the African Middle Stone Age diverged from their Eurasian counterparts, not through isolation alone but via the fading resonance of ancestral voices, allowing novel geometries to flourish.

The Upper Paleolithic explosion, commencing around 50,000 years ago, unveils the crescendo of this resonant orchestration. Homo sapiens, armed with expanded frontal lobes and perhaps a heightened sensitivity to the morphic milieu, unleashed a torrent of innovation. Blade cores flourished in the Aurignacian, yielding slender, parallel removals that birthed burins, end-scrapers, and the exquisite bone points of Sungir. In the Solutrean of Western Europe, laurel-leaf points achieved translucent thinness, their pressure-flaked edges sharper than the finest steel. Across the Bering land bridge in North America, Clovis knappers crafted fluted projectiles that echoed yet transcended Old World antecedents. How to account for such synchronous leaps, from the prismatic blades of Siberia to the microblades of Japan? Our morphic mathematical framework posits a bifurcation in form: new artifacts as superpositions of inherited vectors, novel inventions, and entangled correlations with distant kin. A small fraction of novelty—perhaps twenty percent—blended with the parent, while an entanglement tensor linked specific traits across continents, enabling near-simultaneous emergence of Levallois-like tools in India and the Levant without pedestrian migration.

Delve deeper into the mechanics of guidance, and one discerns a Newtonian mathematical analogy on curved temporal terrain. Acceleration along a path balanced curvature-induced terms resembling the covariant derivative against forces from cognitive limits and resonant nudges. A knapper experimenting with a novel hafting technique felt not random whim but a directed influence, the kernel contracting variations to propel the trajectory. Each perfected tool, upon realization, inscribed an indelible mark—a delta spike in the field, updating it instantaneously across the morphic manifold. This feedback loop accelerated progress; the morphic cloud, ever-enriched, encouraged trials with a score integrated over possibility and transition development spaces, amplifying ventures aligned with historical triumphs.

Consider the transition from Neanderthal Mousterian to Cro-Magnon Aurignacian in the caves of Chauvet and Lascaux. As glacial winds howled outside, Homo sapiens artisans, perhaps intermingling with Neanderthals, sensed the morphic shift. The parent Mousterian field, rich in scrapers and denticulates, faded as the daughter field initialization introduced prismatic cores and organic tools. Noise terms injected variability— a burin spall here, an eyed needle there—while self-coupling strengthened local motifs. Overlap eroded; by 40,000 years ago, the cosine threshold breached, birthing a culture ablaze with symbolic expression. Venus figurines and cave murals, though beyond stone technology, resonated with the same field, their forms echoing the symmetry quests of the prior Acheulean tradition.

Across oceans and ice ages, this resonance wove a global tapestry. In the Americas, post-Clovis diversification yielded Folsom flutes and Dalton points, each a geodesic extension nudged by an ancestral cloud diluted through Beringia. In Australia, the Kimberley points, pressure-flaked to serrated perfection some 5,000 years ago, converged on Old World Solutrean aesthetics despite 50,000 years of separation. Entanglement terms, correlating edge angles and bilateral plans, bridged the unbridgeable, suggesting a non-local influence that transcended migratory paths.

The Neolithic dawn, around 12,000 years ago, marked a resonant pivot as hunter-gatherers settled into agrarian rhythms and animal husbandry. Ground stone axes, polished to a sheen, emerged alongside sickles and querns, their forms guided by a field now layered with microlithic precursors. In the Fertile Crescent, Natufian knappers' composite tools with geometric inserts foreshadowed the composite sickles of Çatalhöyük. The Gaussian weights, with timescales spanning hundreds of millennia, ensured that Paleolithic blade traditions informed Neolithic precision, decay notwithstanding.

Even in the Bronze Age, stone tool technology retained its voice. Obsidian mirrors of Mesoamerica, knapped with prismatic finesse, echoed

Upper Paleolithic blades, while Polynesian adzes carried Acheulean robustness on transoceanic voyages. The morphic cloud, accumulated across 3.3 million years, persisted as an ancestral hum, influencing even metalworkers who retained stone anvils and polishers.

Reflect upon the Holocene acceleration: from the Jomon cord-marked pottery of Japan, adorned with stone tools of exquisite miniaturization, to the Arctic Small Tool tradition's microblades, resonance orchestrated convergence. Daughter fields proliferated—Mississippian celts, Mayan eccentrics—each initializing with parental echoes, mutating under local pressures, and speciating when overlaps waned. In insular realms, isolation amplified noise, yet resonance pierced veils. Tasmanian toolkits, simplified post-separation 10,000 years ago, retained core-flake essentials, the parent's voice decaying yet sufficient to sustain viability. In the cultural crucible of Mesoamerica, Olmec jade celts and Zapotec urns bore the imprint of entangled innovations, superpositions yielding colossal heads and ceremonial bifaces.

The morphic mathematical framework illuminates puzzles long debated: why does stasis predominate in some eras, while bursts enliven others? Morphic geodesics in deep field grooves explain Acheulean longevity—1.5 million years of handaxe fidelity—as field alignment renders any deviation costly. Bifurcations, triggered by climatic pulses or populational fluxes, introduced perturbations, daughter fields drifting until thresholds birthed Mousterian multiplicity or Upper Paleolithic efflorescence.

Anthropocene echoes resound in ethnographic records. Australian Aboriginal craftsmen, demonstrating leaching techniques to anthropologists, replicated technologies extinct for millennia, as the cloud whispered to them through the generations. Inuit ulu makers, curving semiflunar blades, converged on Magdalenian precedents without historical linkage.

Ultimately, the evolution of stone tools transcends the material narrative, revealing a cosmic dance in three-dimensional time. Each flake detached, each core prepared, contributed to a collective memory that guided successors with Gaussian gentleness and kernel precision. Traditions as geodesics, speciations as threshold crossings, innovations as entangled superpositions—these weave a tale of human becoming, where mind resonates with ancestry on morphic manifolds operating beyond clock and calendar.

From the Oldowan dawn to the polished axes of Çayönü, from Levallois laminarity to the elegance of Clovis, the morphic field accumulated, updated, diverged, and reconverged. Knappers, the world's first geometers, navigated curved temporal spaces, their paths minimized by resonant discounts and accelerated by field dynamics. In this poetry of persistence and transformation, stone tools stand eternal, not as relics but as harmonics in a symphony of form conducted by morphic resonance.

This resonance endures. A modern flintknapper, striking a nodule in experimental archaeology, may well feel the pull of Boxgrove's 500,000-year-old symmetry or Gesher Benot Ya'aqov's prepared cores. The cloud, integral over 3.3 million years of triumphs, encourages with undiminished vigor, its Gaussian weights bridging abyssal gaps. In laboratories and hearths, the evolution continues—subtle, profound, eternally human.

Mathematical Appendix: Modeling the Evolution of Stone Tool Technology via Morphic Resonance in Three-Dimensional Time

This appendix adapts a system of equations—originally formulated for biological speciation in a three-dimensional (3D) time manifold—to describe the emergence, stabilization, and non-local transmission of stone tool technologies. Here, "forms" ϕ^i represent knapping techniques, tool morphologies (e.g., handaxe symmetry, Levallois core preparation), or behavioral patterns rather than organismal phenotypes. The morphic field M^i accumulates successful tool-making configurations across hominin populations, enabling resonant influence across space and time via the morphic cloud. The 3D time manifold comprises t^1 (chronological sequence), t^2 (possibility space of tool variants), and t^3 (transition probabilities between variants). Resonance strength decays with temporal distance ds^2, amplifies coherent innovations, and facilitates convergent evolution without direct contact.

Each equation is listed with its symbolic form followed by a brief interpretation in the context of stone tool evolution.

1. **Temporal Metric**

   $$ds^2 = g_{ij}\,\Delta t^i\,\Delta t^j$$

   Measures squared temporal distance in the 3D time manifold between tool-making events (e.g., an Oldowan chopper strike in Africa and a similar event in Asia), determining resonance strength; smaller ds^2 implies stronger non-local influence.

2. **Morphic Field Accumulation**

   $$M^i(t^k) = \int K^i{}_j(t^k, t'^k)\,\phi^j(t'^k)\,d^3t'$$

   Builds the morphic cloud by integrating past tool configurations ϕ^j (e.g., flake angles) weighted by the resonance kernel over all prior events in the 3D time volume.

3. **Gaussian Weight**

$$G(ds^2) = \exp\left(-\frac{ds^2}{2\sigma^2}\right)$$

Applies Gaussian decay to resonance based on temporal separation, with scale σ setting the "memory horizon" (e.g., Acheulean handaxes resonate more strongly within ~1 million years than across 3 million years).

4. **Resonance Kernel**

$$K_{ij}(t^k, t'^k) = G(ds^2)\,\delta_{ij} + \kappa\,\partial_i\partial_j G(ds^2)$$

Combines direct replication of tool forms (Kronecker δ_{ij}) with differential guidance from curvature in possibility space, scaled by coupling κ; favors symmetric handaxes over asymmetric variants.

5. **Geodesic in Morphic Cloud**

$$\exp_p(v) = \gamma(1), \quad \frac{D\gamma}{d\lambda} = M^i(\gamma)\frac{\partial}{\partial t^i}$$

Defines stable tool traditions (e.g., persistent Levallois technique) as geodesics propelled by the morphic field on the manifold, starting from a base point p with tangent v.

6. **Action Functional**

$$S[\gamma] = \int\left[\frac{1}{2}g_{ij}\dot{\gamma}^i\dot{\gamma}^j - V_{\text{gene}}(\gamma) + \kappa\,\delta\phi\cdot M(\gamma)\right]d\tau$$

Quantifies the effective "cost" of a technological trajectory γ (e.g., Oldowan $\rightarrow$ Acheulean transition), minimizing under kinetic terms, cognitive/genetic constraints V_{gene}, and resonant alignment with the cloud.

7. **Resonance Score**

$$\mathcal{R} = \int_{t^2,t^3} \delta\phi_i(t^2,t^3)\,M^i(t^1,t^2,t^3)\,dt^2dt^3$$

Evaluates how well exploratory tool variations $\delta\phi$ (e.g., a novel pressure-flaking trial) align with the established morphic field over possibility and transition dimensions; high $\mathcal{R}$ stabilizes innovations like Upper Paleolithic blades.

8. **Equation of Motion**

$$\frac{D^2\gamma^k}{d\tau^2} + \Gamma^k_{ij}\dot{\gamma}^i\dot{\gamma}^j = -\partial^k V_{\text{gene}} + \kappa\, K^k{}_l\, \delta\phi^l$$

Governs acceleration of technological paths on the curved manifold, balancing Christoffel curvature Γ, cognitive biases V_{gene}, and resonant forces from kernel-applied deviations.

9. **Morphic Field Update**

$$\frac{\partial M^i}{\partial t^1} = \kappa\, \delta^{(3)}(t^k - \gamma^k(\tau))\, \phi^i(\gamma(\tau))$$

Increments the morphic cloud with a delta-function spike when a population finalizes a successful tool form (e.g., a refined microlith), reinforcing it chronologically.

10. **Daughter Field Initialization**

$$M^i_D(t^k_{\text{split}}) = M^i_P(t^k_{\text{split}}) + \epsilon\, \Delta M^i_{\text{mutation}}(t^k_{\text{split}})$$

Initializes a new sub-tradition's morphic field (e.g., regional Levallois variant) from the parent (Middle Paleolithic base) with small perturbation ϵ at the branching point.

11. **Divergence Kernel**

$$K^i_{D\leftarrow Pj} = G_{\text{div}}(ds^2_{PD})\, \delta^i_j\, e^{-\lambda(t^1 - t_{\text{split}})}$$

Models fading influence of the parent tradition on the daughter (e.g., Acheulean echo in early Levallois) via Gaussian form distance and exponential chronological decay λ.

12. **Daughter Field Evolution**

$$\frac{\partial M^i_D}{\partial t^1} = \kappa_D\, K^i_{Dj}\, \phi^j_D + \kappa_{D\leftarrow P}\, K^i_{D\leftarrow Pj}\, \phi^j_P + \eta\, \xi^i(t^k)$$

Evolves the daughter field through self-reinforcement, lingering parent resonance, and stochastic local innovations ξ (e.g., environmental adaptations in blade technology).

13. **Speciation Threshold**

$$\mathcal{O}_{DP}(t^1) = \frac{\int M_D^i M_P^i \, d^3t}{\sqrt{\int M_D^j M_D^j \, d^3t \cdot \int M_P^k M_P^k \, d^3t}} < \theta_c$$

Triggers technological "speciation" (independent industry, e.g., Upper Paleolithic detachment from Middle) when normalized field overlap drops below critical θ_c.

14. **Form Bifurcation**

$$\phi_D^i = \sqrt{1-p}\,\phi_P^i + \sqrt{p}\,\phi_{\text{novel}}^i + \sqrt{p(1-p)}\,\Phi_{\text{entangle}}^{ij}\phi_P^j$$

Constructs a new tool form as a quantum-like superposition: mostly inherited from parent $(1-p)$, partly novel, with entanglement term Φ correlating distant convergent variants (e.g., simultaneous Levallois emergence).

VII

Morphic Resonance in Herd Animals: Collective Instinct in Six-Dimensional Spacetime

Herd animals, such as wild horses, elephants, and buffalo, move through their natural habitats with a unity that transforms individual instincts into a collective symphony. A herd of mustangs thunders across the American plains, swerving as one to outmaneuver a predator, their synchronized strides a testament to shared purpose. Elephants in African savannas navigate to distant waterholes, guided by a matriarch's wisdom that seems to transcend her own experience. Buffalo herds in the Serengeti migrate in vast, coordinated waves, adjusting paths with a precision that defies chaos. These behaviors suggest a hidden intelligence, a force weaving individual actions into a cohesive whole without a central leader. Rupert Sheldrake's theory of morphic resonance offers a compelling explanation, proposing that a species-specific field acts as a collective memory, connecting past and present behaviors across a herd. To deepen this framework, we adapt Dr. Gunther Kletetschka's theory of three dimensional time, turning time itself into a dynamic manifold on which morphic resonance orchestrates herd behavior.

The Enigma of Herd Coordination

The synchronized movements of herd animals are a marvel of the natural world. Wild horses, like those roaming the American West, gallop as a single entity, their turns and acceleration outpacing pursuing wolves with a choreographed grace. Elephants traverse vast savannas in tight-knit family groups, finding waterholes with a precision that suggests a shared map, even for young members new to the journey. Buffalo herds, whether American bison or African cape buffalo, surge across grasslands, shifting direction or speed in unison to evade threats or follow seasonal cues. Schools of fish flow through the water as one, as if animated and directed by a collective intelligence. These actions—evading predators, migrating to grazing grounds, or shielding vulnerable young—point to a deeper mechanism that coordinates the group beyond individual perception.

Traditional explanations rely on sensory signals and social learning. Horses use sharp eyesight and subtle cues, like ear flicks or shifts in stance, to signal alarm or movement. Elephants communicate through low-frequency rumbles, touch, or scent, while buffalo respond to visual cues or the momentum of nearby herd members. Yet, these mechanisms struggle to explain the speed and seamlessness of herd responses. When a lion charges a buffalo herd, the group wheels away instantly, as if guided by a single thought. Elephants find water across generations, even in unfamiliar terrain, suggesting a memory that outlives individuals. Morphic resonance posits that a species-specific field, a non-physical archive of past behaviors, connects herd members across time and space, guiding their actions with an efficiency that complements and surpasses sensory and genetic mechanisms. The theory of three dimensional time offers a framework to understand where this field is located and how it operates, offering a unified stage on which morphic resonance shapes herd dynamics.

Six-Dimensional Spacetime: A Canvas for Resonance

To grasp how morphic resonance guides herd animals, we must reimagine the nature of reality through the theory of three dimensional time. Traditional views treat time as a single, linear progression and space as three dimensions of length, width, and height. Kletetschka expands this into a six-dimensional framework, with three dimensions of time and three of space, unified in a dynamic tapestry. The first time dimension, t1, is the familiar flow of moments, the "now" of experience. The second, t2, represents possibilities, like the potential paths a horse might take when fleeing a predator. The third, t3, governs transitions between these possibilities. Adapted for morphic resonance, t3 orchestrates shifts in animal behavior, like a conductor guiding a herd's response. These time dimensions intertwine with the three spatial dimensions—up-down, left-right, forward-back—forming a six-dimensional canvas where reality unfolds.

In Dr. Kletetschka's perceptive vision, the three spatial dimensions emerge from time through a projection process, like shadows cast from a higher-dimensional object, ensuring that events remain causally linked. Herd animals, like all physical entities, are dynamic patterns—threads or braids—

woven through the resulting six-dimensional fabric, their behaviors tracing paths called worldlines that reflect both individual actions and collective instincts. The morphic field operates within this elegant framework, acting as a collective memory that resonates across time and space, connecting a horse's gallop, an elephant's trek, or a buffalo's charge to the experiences of their ancestors. The morphic field mirrors the non-local connections of quantum systems, where distant entities share information instantly, and the curvature of spacetime, which shapes large-scale interactions.

Morphic Resonance in Herd Animals: Weaving Collective Instinct

Morphic resonance serves as a collective memory that orchestrates herd behavior within three dimensional time. For wild horses, this field holds patterns of flight, such as the precise angles of turn to evade a predator. When a mustang herd senses a wolf, their synchronized gallop suggests a shared knowledge, as if each horse taps into a reservoir of past escapes stored in the field. This resonance, operating across the temporal dimensions, reinforces successful strategies—like choosing open plains over tangled brush—ensuring the herd moves as one. The field acts like an unseen guide, weaving individual movements into a collective dance across the six-dimensional spacetime canvas, in which past and present converge in harmony.

For elephants, the morphic field encodes navigational wisdom, such as the location of waterholes in the vast savanna. A matriarch leads her family to a pond, her decisions informed by a memory that stretches beyond her own life. Young elephants follow, not merely learning but resonating with a field that carries ancestral journeys, as if the paths of past herds whisper directions through the third time dimension. The morphic field, shaped by the interplay of time and space, enables elephants to adapt to droughts or shifting landscapes, their movements a testament to a shared heritage that persists across generations.

Buffalo herds exhibit similar resonance. During migrations across the Serengeti or Great Plains, thousands move in concert, their paths adjusting to avoid predators and find lush grazing lands. When a predator nears, the herd's sudden shift—forming a defensive ring around calves—feels instantaneous, as if each buffalo senses the collective will. The morphic field stores these patterns, from migration routes to defensive formations, amplifying strategies that have ensured survival through countless seasons. In Kletetschka's framework, this field operates non-locally, connecting individuals across spatial and temporal dimensions, allowing a buffalo in the present to resonate with the survival tactics of its ancestors.

The field is dynamic, updated by new experiences within the three dimensional time framework. A horse herd discovering a safer escape route strengthens the field's memory of that path, guiding future flights. An elephant finding a new water source adds this knowledge to the field, benefiting her descendants. Buffalo adapting to a new predator's tactics reinforce defensive patterns, ensuring the field evolves with the herd's needs. This resonance is like a living conversation across time, where past and present actions align to shape survival strategies in the wild.

Individual Roles Within the Collective

While the morphic field coordinates the herd, it also shapes individual animals, aligning their instincts with the group's needs within the six-dimensional canvas. A wild horse's role—leading the charge, guarding the rear, or protecting a foal—emerges from its connection to the field. A young stallion might shift to leading when the field amplifies signals of a predator's approach, its actions guided by a resonant memory that spans the temporal dimensions. This interplay is like a dance, where each horse responds to an unseen rhythm, its movements harmonizing with the herd's flow across space and time.

In elephant herds, individuals take on roles like sentinel, nurturer, or follower, guided by the field's resonance. A young female learns to protect calves by sensing the field's emphasis on maternal instincts, shaped by generations of matriarchs. The field, operating through the third time dimension, nudges her toward actions that support the group, such as standing guard during a rest or aiding a struggling calf. This guidance ensures that each elephant's efforts strengthen the herd's resilience, even in harsh conditions, as their actions resonate across the three dimensional fabric.

Buffalo individuals adapt roles—scout, defender, or core herd member—based on the morphic field's cues. A bull charging a predator may be prompted by a resonant memory of past defenses, while others tighten the herd's formation, their actions aligned through the field's non-local connections. In the morphic framework, this dynamic resembles quantum systems, with individuals exploring multiple roles—like particles in multiple states through superposition—until the field guides them to a specific action, synchronized across space and time, in a manner resembling the collapse of the wave function in quantum mechanics. The natural balance of individual flexibility and group unity enables herds to thrive in unpredictable environments.

The Role of Environment and Feedback

The morphic field is deeply attuned to the environment, weaving in cues like the scent of water, the rustle of a predator, or the texture of terrain. For horses, a rocky gulch prompts the field to favor cautious paths, guiding the herd to safer ground. Elephants respond to drought by resonating with memories of distant waterholes, their field adapting to seasonal shifts within the six-dimensional spacetime framework. Buffalo herds adjust migration routes when grasslands thin, the field incorporating environmental signals to steer them toward areas of abundance. This sensitivity ensures rapid responses to external pressures, maintaining the herd's efficiency.

Feedback within the herd refines the field. A horse leading a successful escape strengthens the field's memory of that route, influencing future movements. An elephant matriarch's choice of a reliable water source reinforces that path, guiding her successors. Buffalo forming a tight defensive ring against a lion amplify this strategy, ensuring it persists across generations. In the morphic resonance model, this feedback loop operates across temporal dimensions, creating a self-reinforcing system where individual actions enhance collective patterns, blending environmental awareness with herd wisdom to sustain the group's cohesion and adaptability.

Philosophical Horizons

Morphic resonance, embedded in the context of three dimensional time, portrays herd animals as emergent entities, unified by a field that transcends physical boundaries. This field, operating across time and space, suggests a universe where memory is a fundamental force, linking individuals in a kind of shared consciousness. This vision echoes ancient mystic philosophers who saw nature as an interconnected whole, while still inviting scientific exploration through testable predictions about such matters as coordination speed or inherited behaviors.

This perspective extends beyond the wild, suggesting that human societies and political systems might reflect similar resonant fields and, indeed, be influenced by the mathematics of morphic resonance. Collective human entities, such as nations, states, and parties, could mirror the herd's dynamics. By viewing herds of horses, elephants, and buffalo as expressions of morphic resonance within a three dimensional canvas of time, we glimpse a world where memory, time, and instinct intertwine, crafting a collective dance that sustains life in the wild with elegance and unity, while also underlying the development of human societies and history.

VIII

Morphic Fields and Early Tribal Societies

The oldest form of human political organization, the tribe, reflects the coordinated movements observed in herds of horses, elephants, and buffalo across the great open landscapes of the world, where individual motions dissolve so completely into a single collective flow that the boundary between one organism and many often seems to vanish. A tribal band, resembling a herd of mustangs sweeping in perfect synchrony over the North American grasslands, advances with unified intent during the great communal hunts, the seasonal migrations, or in sudden defence against danger. Elephants cross the sun-scorched savannas of Africa drawing upon an accumulated group knowledge that appears to stretch far beyond the personal memory of any living member, while buffalo move in vast rhythmic surges directed by an invisible guiding presence sustained within the morphic field.

The theory of morphic resonance, which proposes that nature is regulated by a kind of inherent memory, describes species-specific and group-specific fields that function as collective memories, connecting behavioral patterns of the past with those of the present across individuals, generations, and considerable distances. The six dimensional spacetime framework discussed frequently in this book supplies the deeper underlying structure for such resonance, with worldlines marking the dynamic trajectories of behavioral habits through a multidimensional manifold whose three temporal dimensions are primary and whose spatial dimensions emerge as secondary projections.

Comparisons arise naturally and repeatedly between the morphic dynamics of human tribes and the morphic dynamics of animal herds, particularly in the way traditional leaders emerge through repeated demonstrations of physical strength, courage, and practical wisdom, and, in the human case alone, through the additional and decisive influence of ancestral descent stretching back across dozens or hundreds of generations. Such morphic processes, operating outside the conventional channels of genetic inheritance and cultural teaching, may account for the historical emergence and astonishing persistence of aristocracy—the deep, almost instinctive sense, found in nearly every part of the inhabited world, that particular lineages possess an inherent, field-amplified capacity for governance, counsel, and sacred authority.

The Unity of Tribes and Herds: A Common Morphic Ground

The first unmistakably human tribes appeared at least seventy thousand years ago, perhaps considerably earlier, binding small bands of hunter-gatherers—typically numbering between twenty and one hundred and fifty persons—into remarkably coherent wholes whose survival depended upon an almost mystical level of coordination. Groups such as the San of the Kalahari, the Hadza of Tanzania, the Mbuti of the Ituri forest, or the Australian Pintupi maintained fluid yet unbreakable solidarity while coordinating complex hunting drives, fire management, seasonal movement, and collective defence without written rules and laws.

Herd animals manifest comparable cohesion in the present day, and the parallels are too striking to be dismissed as mere analogy. Wild horses in the American West or the Camargue of southern France shift direction simultaneously when threatened, turning as one organism with a precision that defies explanation by visual cues alone. Elephants in the savannas of Kenya, Botswana, or Sri Lanka locate distant, seasonal water sources through knowledge that appears to exceed the lifespan of any single matriarch, knowledge that is transmitted with uncanny fidelity even to young calves who have never travelled the route before. Buffalo on the Serengeti plains or the floodplains of the Okavango form vast migrating columns containing tens of thousands of individuals, or, in an instant, contract into perfect defensive circles around the vulnerable young, responding with a rapidity and unanimity that transcend ordinary sensory signalling and learned imitation.

Conventional analysis attributes such unity to social bonding, oral transmission, ritual reinforcement, and subtle bodily cues in tribes, or to ear positions, tail flicks, and infrasonic rumbles in animals. Yet the instantaneous quality of coordinated response—whether San hunters closing upon a kudu in absolute silence or an entire buffalo herd wheeling away from a lion pride in the space of a heartbeat—points toward a deeper connective medium that operates more rapidly than light could travel through ordinary three-dimensional space if the signals were purely local. Morphic resonance posits the existence of non-material fields that preserve and transmit behavioural habits across time and distance by a process of resonant tuning rather than energetic transfer, influencing both human and animal groups with equal facility. Within the six-dimensional spacetime manifold, the worldlines of individuals interweave like threads in a loom, revealing structural parallels between the social organizations of humans and those of large herbivores, including the emergence of leadership through proven capability in the moment and, in human societies alone, the additional and cumulative weighting given to ancestral continuity across the centuries.

Six-Dimensional Spacetime: The Underlying Fabric

Reality conceived in six dimensions presents three primary temporal dimensions from which the familiar three spatial dimensions derive as secondary projections or shadows, thereby integrating the smooth curvature described by General Relativity with the probabilistic, non-local fields of quantum theory. The first temporal dimension corresponds to ordinary sequential experience—the arrow of clock time we all inhabit. The second temporal dimension encompasses branching possibilities: the hunter's choice of which trail to follow at the fork, the horse's momentary impulse either to graze or to flee, the multitude of unrealised futures that quantum theory treats as superposed states. The third temporal dimension governs the selection and orchestration of transitions among those possibilities, guiding sudden shifts in collective behaviour that appear instantaneous from the perspective of ordinary time.

The three spatial dimensions—height, breadth, depth—emerge mathematically as projections from this richer temporal manifold, much as a two-dimensional shadow is cast by a three-dimensional object. Physical entities, whether human hunters moving silently through acacia scrub or migrating buffalo crossing a crocodile-infested river, trace worldlines across this six-dimensional continuum. Gravitational curvature, quantum superposition, and quantum entanglement find common expression when spatial dimensions are understood as derivative rather than fundamental. Morphic fields operate upon this deeper temporal canvas, enabling past behavioral configurations—ritual dances performed around a campfire seventy generations ago, or the defensive formations adopted by buffalo ancestors during the Pleistocene—to resonate along present worldlines, producing effects that resemble quantum non-locality while remaining fully consistent with the causal structure of an enlarged spacetime.

Morphic Resonance among Human Tribes: Memory, Coordination, and Inherited Authority

Within early human tribes, morphic fields served as distributed collective memories that orchestrated group action across all three temporal layers of the six-dimensional manifold. Even today, San hunters pursue antelope with movements so finely attuned that an external observer might suppose a single directing intelligence distributed among twenty separate bodies. Patterns of successful hunting, fire-making, seasonal movement, conflict mediation, and trance dancing persist within the field as habitual grooves or attractors, resonating forward through the second and third temporal dimensions to influence present participants with a gentle but irresistible pressure toward coherence. Camp relocation, the division of meat after a kill, the simultaneous decision of fifty individuals to break camp at dawn—all proceed with a smoothness suggesting direct access to an accumulated archive of prior solutions that no single member could possibly hold in personal memory.

Leadership commonly crystallises around demonstrated strength and practical wisdom: success in bringing down large game, courage in the face of raiders, skill in healing or mediation. The worldline of such a leader becomes temporarily privileged within the field, resonating more powerfully with memories of past effectiveness stored in the collective archive. Yet human tribes differ profoundly from animal herds in the pronounced role accorded to ancestry across dozens, sometimes hundreds, of generations. Descent from celebrated hunters, shamans, or peacemakers confers enhanced resonance; the field appears to recognise and amplify the influence of certain lineages, embedding their past achievements so deeply into the living memory of the group that descendants are born into an atmosphere already charged with expectation and deference.

Among the Yolngu of Arnhem Land, for example, the right to paint certain songline designs or to lead particular ceremonies belongs exclusively to clans whose ancestors first received those designs in the Dreaming; the morphic field itself seems to guard the boundary, making unauthorised performance feel not merely improper but faintly impossible. Among the Nuer of South Sudan, aristocratic "leopard-skin chiefs" trace their authority to founding ancestors whose peacemaking feats still vibrate in the field centuries later. Successful actions by members of such lineages strengthen the field still further, deepening the impression that particular bloodlines are naturally fitted to guide, to mediate between human and spirit worlds, and to speak with authority in council. Over generations this cumulative process nurtures and stabilises the concept of aristocracy—an intuition, shared from the Celtic túatha to the Polynesian ariki, that governance belongs by nature to specific families whose worldlines carry the strongest and clearest signal within the six-dimensional resonant spacetime fabric.

Morphic Resonance in Animal Herds: Pure Instinctive Coherence

Parallel processes, stripped of cultural elaboration, appear in the great herbivore herds. Wild horses execute sweeping turns that evade wolf or lion with balletic precision, drawing upon field-stored patterns of escape refined across hundreds of thousands of years. Elephants recall the locations of waterholes visited decades earlier by ancestors long dead, transmitting that knowledge through the matriarch with a fidelity that genetic inheritance alone cannot explain. Cape buffalo adjust formation instantaneously when predators approach, the field reinforcing the circular defence or the headlong charge across countless generations and continents.

Leadership emerges through direct, unambiguous trials: a stallion maintains his position by prevailing in physical combat with challengers, an elephant matriarch by accumulated navigational success and social intelligence, a dominant buffalo bull by raw power repeatedly displayed in defence of the herd. Their worldlines temporarily dominate the field, yet authority remains contingent upon continuing performance; a younger, stronger challenger can displace a leader the moment superiority is demonstrated. Offspring may inherit proximity to leadership—the daughter of an elephant matriarch walks immediately behind her mother and learns the routes—but only present capability, demonstrated anew in each season, secures the role. The field in herds lacks the cultural amplification of lineage found in human groups; memory remains tied strictly to survival utility rather than to narrative tradition or sacred history.

Comparison of Tribal and Herd Morphic Dynamics

Striking convergences mark the morphic organisation of tribes and herds within six-dimensional spacetime. Collective memory guides both systems with equal elegance: a tribal hunting party moving through tall grass and a fleeing horse band display identical dissolution of individual into group intention, as though a single extended organism had momentarily borrowed many bodies. Navigation to distant water by elephants mirrors tribal movement to new seasonal camps, each drawing upon trans-generational resonance facilitated by non-local connections across the second and third temporal dimensions.

Leadership crystallises similarly around proven strength and effectiveness: the skilled tribal tracker who brings meat to the fire and the victorious stallion who keeps the herd intact both occupy amplified positions within their respective fields, positions that attract followers by the sheer clarity of the resonant signal they emit. Yet the human field uniquely sustains ancestral weighting across centuries. Maasai moran from renowned warrior families carry enhanced authority before they have killed their first lion, as though past victories of their grandfathers and great-grandfathers continue to vibrate within the morphic medium, lending weight to spear and voice alike. Herd fields, lacking storytelling, songlines, and ancestral shrines, grant no comparable hereditary privilege; dominance remains perpetually contestable and usually short-lived.

Aristocracy as a Morphic Phenomenon

Morphic resonance offers a coherent and testable explanation for the persistence of aristocratic forms in human societies from the earliest historical horizon to the present. Each successful action by a member of a revered lineage—each decisive hunt, each wise judgement, each ritual performance that brings rain or heals division—intensifies the field's memory of that lineage's competence, conferring upon descendants a stronger initial resonance, a kind of morphic head-start. Iroquois clan mothers drawn from particular families exercised authority reinforced by centuries of remembered wise judgement. The field itself appears to select for continuity of governance within certain lines, generating the widespread and remarkably stable conviction that noble blood carries intrinsic fitness to rule.

Animal herds, by contrast, exhibit no equivalent hereditary stabilisation; leadership passes according to immediate demonstration of ability, unencumbered by ancestral narrative. An ageing matriarch may be shouldered aside by a younger sister the moment her memory falters on the migration route; a buffalo bull defeated in combat simply drifts to the periphery of the herd. The difference arises from the presence in humans of language, ritual, and cumulative cultural memory, which allow the morphic field to store and amplify lineage-specific patterns across far longer stretches of time than mere survival utility would require.

Environmental Attunement and Field Evolution

The morphic fields of both tribes and herds remain exquisitely responsive to landscape, season, and ecological change. Diminishing game leads a tribe to new territories, just as prolonged drought redirects elephant migrations, the field incorporating and transmitting these adjustments with remarkable speed. Successful outcomes—abundant forage, safe passage, the birth of healthy calves—strengthen corresponding patterns, creating positive feedback loops that refine collective behaviour across all three temporal dimensions of the six-dimensional manifold. In human groups, stories of ancestral adaptation further reinforce lineage authority, whereas animal fields evolve solely through the impersonal criterion of survival utility.

Ecological and Evolutionary Consequences

Morphic resonance accounts for the otherwise puzzling adaptability displayed by hunter-gatherer tribes and large herbivore herds alike. The San persisted for millennia in one of the harshest deserts on earth, and elephants continue to navigate increasingly fragmented habitats, both depending upon morphic fields that preserve and update navigational and tactical knowledge far faster than genetic change alone could achieve. Convergent social forms—similar spear points appearing simultaneously among separated tribes, comparable defensive formations in zebras, wildebeest, and buffalo—suggest access to shared species or group fields that operate across continents and millennia. The hereditary emphasis in human aristocracy may represent an evolutionary stabilization of leadership resonance that confers decisive advantages in times of crisis, when rapid, unified response under trusted command can mean the difference between survival and extinction.

Philosophical and Scientific Perspectives

Within a six-dimensional framework, morphic resonance reveals tribes and herds as emergent wholes sustained by fields that transcend local causality and ordinary energetic signalling. Human societies extend these fields through language, ritual, and sacred history, generating stable governance structures and the aristocratic principle that has shaped civilisation from Çatalhöyük to the present. Animal herds manifest a purer, less elaborated resonance centred exclusively on immediate survival. The resulting vision of living systems woven together by invisible memories moving through a multidimensional spacetime links rigorous scientific description with the most ancient intuitions of interconnected being, while inviting empirical exploration of such key phenomena as coordination speeds, inheritance patterns of authority, and field-mediated learning across isolated populations. From the first flickering fires of tribal councils under Pleistocene stars to the sweeping seasonal migrations of the Serengeti that have continued unchanged for millions of years, nature reveals itself as a vast resonant dance in which memory, time, ancestry, and collective form intertwine to produce the graceful, enduring coherence that makes complex life possible.

IX

Morphic Resonance and the Emergence of Early Kingdoms

In the tapestry of human history, the transition from tribal societies to early kingdoms marks a profound shift, as small, egalitarian bands coalesced into hierarchical structures under centralized authority. This transformation, spanning millennia, reflects not only social and political evolution but also a deeper interplay of collective memory and individual ingenuity. Rupert Sheldrake's theory of morphic resonance offers a lens to understand this process, positing a species-specific field—a morphic cloud—that stores and transmits behavioral patterns across generations. Within this framework, the concept of three-dimensional time provides a dynamic canvas for this morphic cloud, where past, present, and potential futures intertwine to shape societal development. This essay explores how morphic resonance facilitated the cohesion of tribal societies and their eventual transformation into kingdoms, examines the tension between the morphic cloud's traditional memory and human innovation, and employs an adapted version of the three-dimensional time framework to model this interplay. The same mathematical framework explored in preceding chapters can easily be adapted to model the transition from tribes to early kingdoms.

The Morphic Cloud and Tribal Cohesion

Tribal societies, emerging tens of thousands of years ago, were initially small, kinship-based groups of 20 to 150 individuals, bound by shared rituals, oral traditions, and survival strategies. The San Bushmen of the Kalahari, for instance, moved as one across arid landscapes, their hunters tracking antelope with an almost preternatural unity. This cohesion, akin to a flock of starlings weaving through the sky, suggests a collective intelligence beyond individual cognition. Sheldrake's morphic resonance proposes that a morphic cloud—a non-physical field—acts as a repository of collective memory, linking individuals across time and space. For early tribes, this cloud encoded patterns of hunting, foraging, and ritual, resonating through shared practices to ensure group survival.

Imagine a tribal fire, its embers glowing under a starlit sky. As elders recount tales of successful hunts or sacred rites, the morphic cloud absorbs these patterns, transmitting them to the young like whispers carried on the wind. A hunter's precise spear throw or a healer's herbal remedy becomes part of the cloud, accessible to others in the tribe, even across generations. This resonance fosters unity without written laws, as each member draws on the collective memory to navigate challenges, from drought to conflict. The cloud, in essence, is the tribe's invisible elder, guiding actions through a shared archive of experience.

In the three-dimensional time framework, this morphic cloud operates within a dynamic temporal structure. The first time dimension, akin to linear progression, tracks the sequence of events—daily hunts, seasonal migrations. The second dimension encompasses possibilities, such as a hunter choosing between paths or a tribe debating a new campsite. The third dimension orchestrates transitions between these possibilities, selecting optimal outcomes based on past successes encoded in the cloud. This framework positions the morphic cloud as a multidimensional archive, where tribal behaviors are worldlines—dynamic threads weaving through time, shaped by resonance with ancestral patterns.

From Tribes to Early Kingdoms: The Role of Morphic Resonance

The transition from tribes to early kingdoms, evident in societies like the Sumerians of Mesopotamia or the early Yoruba kingdoms of West Africa, involved the consolidation of power, the rise of permanent settlements, and the emergence of centralized leadership. Morphic resonance played a pivotal role in this process by preserving and amplifying the patterns that enabled larger, more complex societies. In a tribe, the morphic cloud reinforced egalitarian practices, such as communal resource sharing, as seen among the Aboriginal peoples of Australia, where collective rituals ensured group cohesion. As populations grew and agriculture stabilized food supplies, the cloud began to encode new patterns: land division, trade networks, and rudimentary governance.

Consider a tribal community on the cusp of becoming a kingdom. The morphic cloud, rich with memories of successful hunts and shared feasts, now incorporates patterns of surplus storage and territorial defense. A leader, once chosen for hunting prowess, now oversees grain distribution or resolves disputes over land. The cloud resonates with these evolving roles, stabilizing them across generations. In Sumerian city-states like Uruk around 3000 BCE, rulers emerged as priest-kings, their authority bolstered by rituals that resonated with the cloud's memory of spiritual leadership. This resonance lent legitimacy to nascent hierarchies, as the collective memory of revered leaders—shamans, warriors, or judges—crystallized into the concept of kingship.

The three-dimensional time manifold e,bodies and encodes this transition. The first dimension records the historical shift from nomadic tribes to settled communities. The second explores possibilities, such as adopting agriculture or forming alliances with neighboring groups. The third dimension orchestrates these transitions, selecting patterns that enhance survival—settlements over nomadism, kingship and aristocracy over egalitarianism. The morphic cloud, operating within this temporal framework, acts as a bridge between past and future, ensuring that tribal unity evolves into the structured cohesion of kingdoms. The worldlines of individual leaders, once transient, become enduring threads, their resonance amplified by the cloud's memory of governance. In this way, the ideas of kingship and aristocracy become preferred geodesic pathways engraved in the field of morphic resonance.

The Morphic Cloud and the Force of Tradition

The morphic cloud, while adaptive, is inherently conservative, embodying the force of tradition. It prioritizes patterns that have proven successful, reinforcing behaviors that ensure group survival. In a tribal context, this might mean adhering to time-tested hunting routes or ritual practices, as seen among the Maasai, whose cattle-herding traditions remain steadfast across centuries. As tribes evolved into kingdoms, the cloud preserved hierarchical structures, such as divine kingship in ancient Egypt, where pharaohs were seen as living embodiments of the god Osiris, their authority rooted in the cloud's memory of sacred leadership.

This conservative force manifests as a collective inertia, resisting change to maintain stability. A village elder's insistence on traditional planting methods or a king's adherence to ancestral laws reflects the cloud's resonance with past successes. In this framework, the first time dimension anchors these patterns, embedding them as stable worldlines. The third dimension reinforces their persistence, selecting against deviations that might disrupt the group. For instance, in the early Yoruba kingdoms, the Oba's authority was tied to ancestral spirits, their resonance in the cloud ensuring that governance remained tied to tradition, even as trade and urbanization introduced new pressures.

Yet, this conservatism can stifle progress. A tribe clinging to outdated hunting grounds risks starvation in a changing environment, just as a kingdom enforcing rigid laws may falter against innovative rivals. The morphic cloud, while a repository of wisdom, can become a chain, binding societies to the past in a harmful manner. This tension sets the stage for the interplay with human innovation, the spark that drives change within the constraints of tradition.

Human Innovation: The Catalyst of Change

Human innovation—whether a new tool, a novel social structure, or a revolutionary idea—challenges the morphic cloud's conservative pull. In tribal societies, innovation often arose from necessity: a hunter crafting a sharper spear or a healer discovering a new plant's medicinal properties. These innovations, if successful, were absorbed into the cloud, reshaping its patterns. For example, the invention of the atlatl, a spear-throwing device, revolutionized hunting among early humans, its adoption resonating through the cloud to become a standard practice. In the mathematics of morphic resonance, such innovations can be modeled with the equations of speciation, described in Chapter Two.

As tribes grew into kingdoms, innovation became more complex, encompassing irrigation systems, writing, and centralized taxation. In Mesopotamia, the development of cuneiform writing around 3400 BCE transformed record-keeping, enabling kings to manage vast resources. Such innovations disrupted the cloud's traditional patterns, forcing it to adapt. A scribe's new script or a king's novel law resonated as a new worldline, challenging the inertia of ancestral practices. In our analytical framework, the second time dimension is critical here, representing the exploration of possibilities—new tools, new governance models—that deviate from the cloud's established ancestral patterns. The third dimension evaluates these innovations, integrating those that enhance survival into the collective memory.

Innovation, however, faces resistance. The morphic cloud, steeped in tradition, may reject novel practices that lack immediate success. A tribe skeptical of a new farming technique or a kingdom wary of foreign alliances reflects this resistance, as the cloud prioritizes proven patterns. Yet, persistent innovation, backed by tangible benefits, can shift the cloud's resonance. The adoption of bronze tools in the Indus Valley civilization, for instance, overcame initial skepticism, their utility reshaping the cloud to favor metalworking over stone tools. In other cultures, such as the civilizations of Mesoamerica, metal working remained a marginal technology, confined to ritual items and ornaments, while stone tool technology was preserved and refined to heights never reached elsewhere.

The Interplay of Tradition and Innovation

The evolution from tribes to kingdoms is a dance between the morphic cloud's tradition and human innovation, each shaping the other within the matrix of three-dimensional time. Tradition, embodied in the cloud, provides stability, ensuring that tribes survive through shared practices and that kingdoms maintain order through established hierarchies. Innovation, driven by individuals or small groups, introduces change, pushing societies toward greater complexity. This interplay is not a simple tug-of-war but a dynamic mathematical equilibrium, in which tradition grounds innovation, and innovation refreshes tradition.

Consider the rise of the Zulu kingdom in the early 19th century under the monarch Shaka. The morphic cloud preserved tribal traditions of cattle herding and communal warfare, but Shaka's innovations—short stabbing spears, disciplined military formations, and centralized authority—revolutionized Zulu warfare and society. These changes faced resistance, as warriors clung to traditional long spears, but their success in battle resonated through the cloud, redefining its patterns. In our analytical framework, Shaka's innovations explored the second time dimension, testing new possibilities, while the third dimension integrated them, aligning the cloud with a new model of kingship.

This interplay is evident in other transitions. In ancient China, the shift from tribal confederacies to the Shang dynasty involved innovations like bronze metallurgy and oracle bone writing, which disrupted traditional patterns but were absorbed into the cloud as they proved effective. The cloud's conservative force ensured gradual adoption, preventing societal collapse, while innovation drove the complexity needed for the creation of kingdoms. The three-dimensional time framework captures this process: the first dimension records the historical adoption of new practices, the second explores their potential, and the third orchestrates their integration, balancing tradition and change.

Three-Dimensional Time as the Framework of the Morphic Cloud

Three-dimensional time provides a rigorous framework for understanding the morphic cloud's role in societal evolution. Traditional linear time (the first time dimension, t1) tracks the progression of events—such as tribal hunts, kingdom formation, and the adoption of writing. The second dimension, representing possibilities, allows individuals to experiment with innovations, such as new tools or governance structures. The third dimension, the orchestrator, evaluates these possibilities against the cloud's memory, selecting those that enhance survival and integrating them into the collective field.

The morphic cloud exists within this multidimensional temporal canvas, its resonance shaped by the interplay of these temporal dimensions. Tradition, rooted in the first dimension, stabilizes the cloud, preserving patterns like tribal rituals or kingly authority. Innovation, explored in the second dimension, challenges these patterns, introducing new worldlines that may disrupt the cloud's equilibrium. The third dimension resolves this tension, determining whether an innovation—say, a new irrigation system—resonates strongly enough to reshape the cloud. This framework positions the cloud as a dynamic entity, not a static archive, capable of adapting to change while maintaining continuity.

For example, in the transition to kingdoms, the morphic cloud's resonance with tribal leadership patterns (first dimension) was tested by innovations like permanent settlements or taxation (second dimension). The third dimension evaluated these changes, integrating successful ones into the cloud, such as the bureaucratic systems of ancient Egypt. This process ensured that kingdoms retained the cohesion of tribes while embracing the complexity needed for larger societies. It is of great interest that the same equations used in Chapter Two to describe speciation and in Chapter Four to model the emergence of Homo sapiens from Heidelberensis can also describe the emergence of early kingdoms from tribal societies. The emergence of a new human socio-political form from a precious type of society is thus a form of societal speciation.

When Tradition Yields to Innovation

The tension between tradition and innovation reaches a critical point where the morphic cloud's conservative force yields to new ideas, enabling societal leaps like the formation of kingdoms. Drawing on the aforementioned equations, we can model this transition as a balance between the resonant strength of traditional patterns and the disruptive energy of innovation. The morphic cloud's influence can be conceptualized as a potential field, where tradition exerts a stabilizing force, and innovation introduces a perturbative force that seeks to realign the field.

In this model, tradition is represented by the cumulative resonance of past behaviors, stored in the cloud and reinforced across the first time dimension. Innovation, arising in the second dimension, introduces a force proportional to its novelty and impact, evaluated by the third dimension. The tipping point occurs when the innovative force exceeds the cloud's traditional resonance, reshaping its patterns. This can be expressed mathematically, with a system of equations that balances the morphic resonance potential against the innovation potential (see annexes for Chapters Two and Four).

Qualitatively, this tipping point is evident in historical transitions. In ancient Mesopotamia, the morphic cloud preserved tribal patterns of communal labor, but innovations like irrigation and cuneiform writing introduced new possibilities. When these innovations proved successful—enabling surplus food and record-keeping—they overcame traditional resistance, resonating through the cloud to support the emergence of kingdoms. The third time dimension facilitated this shift, selecting innovations that aligned with survival needs, such as centralized governance, over outdated tribal practices.

The interplay of the morphic cloud and innovation has ecological and cultural ramifications. Ecologically, the cloud's resonance with environmental patterns—such as seasonal migrations or water sources—ensured tribal survival, while innovations like agriculture reshaped landscapes, enabling kingdoms to thrive. Culturally, the cloud preserved myths and rituals, as seen in the Epic of Gilgamesh, which reinforced

Sumerian kingship, while innovations like legal codes expanded cultural expression.

In the framework of three dimensional time, these implications unfold across temporal dimensions. The first dimension preserves ecological and cultural patterns, the second explores innovations like farming or literacy, and the third integrates them, ensuring adaptability. The morphic cloud, as a repository of both ecological and cultural memory, bridges tribes and kingdoms, its resonance evolving with each successful innovation.

Morphic resonance, framed in three-dimensional time, invites profound questions about human progress. Philosophically, it suggests a collective consciousness, in which societies are shaped by an invisible memory that transcends individuals. Scientifically, it challenges outdated "materialist" paradigms, proposing a temporal field that links past and present behaviors. The transition from tribes to kingdoms, viewed through this lens, is a testament to humanity's ability to balance tradition and innovation. The morphic cloud, operating within the three dimensional temporal framework, provides the continuity needed for cohesion, while innovation drives the complexity of kingdoms. This interplay, modeled mathematically, reveals a universe where memory, time, and creativity intertwine, crafting societies that endure and evolve with grace and resilience.

X

The Morphic Cloud of Pharaonic Egypt

The civilization of Pharaonic Egypt endured along the Nile for more than three thousand years, from the unification around 3100 BCE to the death of Cleopatra VII in 30 BCE. Few human societies have displayed such sustained coherence of form. The same motifs recur across millennia: the pharaoh smiting enemies, the falcon Horus, the scarab pushing the sun-disc, the ankh held to the nostrils of the dead. The same rituals are performed with only gradual modification; the same gods are worshipped under the same names; the same principle of Ma'at—cosmic order, truth, reciprocity—remains the explicit foundation of kingship, law, and moral discourse. Hieroglyphic writing keeps its essential grammar and early vocabulary from the Early Dynastic period to the Ptolemaic and Roman eras. Mummification techniques alter in detail yet never abandon the core intention of preserving the body for the ka and ba spirits of the dead.

Conventional explanations of Egypt's astonishing continuity invoke geographical isolation, a powerful priesthood, the inertia of a scribal bureaucracy, and the annual flood that enforced a stable agricultural rhythm. These factors are real, yet they describe only local, energetic mechanisms of transmission—parent to child, master to apprentice, temple to temple. Something more pervasive, something non-local and non-energetic, seems at work, something that made certain thoughts, gestures, proportions, and practices feel self-evidently "Egyptian" to people separated by centuries and by hundreds of kilometres of river valley.

The hypothesis of morphic resonance, now expressed with mathematical precision through a three-dimensional cultural time framework (fully derived in the Mathematical Annex at the end of this essay), offers the explanatory bridge. Cultural evolution does not unfold along the single arrow of clock time. It moves within a three dimensional temporal manifold whose three coordinates represent distinct dimensions of cultural organization:

- c1: political unification and centralization
- c2: monumental and institutional elaboration
- c3: religious and symbolic consolidation

The functional distance between any two cultural states is given not by calendar years but by the cultural temporal metric (Equation 1 in the Annex). Only states that are genuinely homologous across these three dimensions have a small ds^2 factor and can therefore resonate strongly. The Gaussian weight (Equation 3) ensures that morphic influence decays smoothly with increasing cultural dissimilarity, favoring near-identical historical phases while still allowing graded influence from less exact matches.

The Emergence of the Field: Predynastic to Early Dynastic

During the Predynastic centuries the Nile valley was occupied by farming communities whose material culture displayed rapidly increasing similarity. Pottery of the Naqada II period appears with closely comparable motifs from Upper Egypt to the Delta. Slate palettes, mace-heads, and ivory carvings bear scenes of animals arranged in rows, ships with high prows, and figures wearing the white or red crown—motifs that will persist for thousands of years. These resemblances cannot be explained solely by trade or political control from a single centre; they arise almost simultaneously over great distances.

At this stage a collective morphic field begins to take shape. As more communities bury their dead in contracted positions with similar pots, carve combs with bird motifs, and brew beer in comparable vats, the cumulative past of all such acts strengthens the field (Equation 2). New communities find it easier to adopt the same practices even in the absence of direct contact. The resonance kernel (Equation 4) combines a direct pull toward ancestral patterns with a second-order term that smooths paths along historically frequent routes, producing the characteristic canalisation we observe.

Around 3100 BCE the political unification under Narmer dramatically intensifies the process. A single kingship now overlays the entire cultivable length of the Nile. The king is Horus; the king is the guarantor of Ma'at; the king is the spiritual channel through whom the annual life-giving inundation of the Nile itself is secured. These conceptions, once enacted as rituals at Abydos, Hierakonpolis, and Memphis, immediately begin to resonate throughout the new realm. The societal trajectory follows the geodesic defined by Equation 5, driven by the morphic field vector M^i(γ) that already points strongly in the direction of divine kingship because of the dense clustering of earlier local chiefdoms in the c1 dimension. Scribes in distant provinces do not need detailed instructions; the pattern is present in the field, and local elites tune themselves to it with remarkable speed.

Old Kingdom: Crystallisation of Monumental Form

By the Third Dynasty the morphic field has grown strong enough to sustain patterns of extraordinary complexity and scale. The Step Pyramid of Djoser at Saqqara marks the first appearance of a new architectural habit: the translation of palatial and funerary forms into cut stone on a colossal scale. Imhotep's innovation does not arise in a vacuum. The field has been prepared by centuries of mastaba-building in mud-brick; the cumulative past of rectangular tombs with niched façades exerts a powerful tuning influence. Once the leap to stone has been made, the pattern resonates so powerfully that within a century the true smooth-sided pyramid appears at Meidum and Dahshur, and then, with the Fourth Dynasty, reaches its classical form at Giza.

The precision of orientation, the consistency of proportions, the repeated use of the 14/11 slope or 5½-palm seked—these features are not enforced solely by written plans, of which few have survived, or by travelling master builders. The action functional (Equation 6) is dramatically lowered whenever a trajectory stays close to the cumulative past. The term equation rewards alignment with historical averages; building in cut stone, orienting to true north within minutes of arc, and conceiving of the king's burial as an ascent along solar rays cease to be technical choices and become the path of least resistance in cultural phase space.

Meanwhile other patterns crystallise with equal inevitability: the solar temples of the Fifth Dynasty, the canonical proportions of human figures in relief (eighteen fists to the hairline), the daily ritual performed in every temple from Elephantine to the Delta, the seventy-day mummification process, the Opening of the Mouth ceremony. Each repetition strengthens the corresponding sub-field; each sub-field contributes to the larger morphic cloud of Egyptian culture as a whole. The resonance score (Equation 7) for any significant deviation from these patterns has become vanishingly small.

Collapse and Reorganisation: The First Intermediate Period

Around 2181 BCE the Old Kingdom disintegrates. Central authority collapses; the inundation fails repeatedly; provincial nomarchs rule from their own cities. A superficial reading might suggest that the morphic field has been shattered. Yet the remarkable fact is how much survives. Local rulers continue to call themselves "Horus," "Son of Ra," and "Beloved of Ptah." They continue to be buried beneath small pyramids or in rock-cut tombs decorated with scenes of daily life drawn from Old Kingdom models. Hieroglyphic inscriptions retain the same grammar and idiom. The god Osiris, whose cult had been growing in importance, now spreads rapidly, but within the existing framework of royal resurrection and legitimacy.

The field has not broken; it has become decentralised and more flexible. Cultural patterns that were formerly sustained by the court at Memphis now draw strength from hundreds of local centres simultaneously. Only trajectories that successfully reach maturity stamp their final institutionalised pattern back along the entire path (Equation 9). Failed micro-kingdoms contribute almost nothing to the updating the morphia field; successful ones—those that still "felt" Egyptian—continue to reinforce the deep attractor basin.

When Mentuhotep II reunites the country around 2055 BCE, the resonance score for a return to Memphite-style kingship is overwhelmingly higher than any alternative. The equation of motion (Equation 8) accelerates the culture back down the pre-existing morphic groove with astonishing ease. Middle Kingdom art and literature consciously imitate Old Kingdom models; the pyramid form is revived, albeit on a reduced scale. The field has demonstrated its resilience: even prolonged political fragmentation cannot erase centuries of cumulative morphic resonance.

New Kingdom: Elastic Expansion and the Amarna Excursion

The expulsion of the Hyksos around 1550 BCE marks another threshold. Chariots, composite bows, scale armour, and new metallurgical techniques enter the Egyptian repertoire. A purely mechanistic account would stress military necessity and conscious imitation of the military technology of rival states. Yet the speed and completeness with which these elements are absorbed suggest the operation of the moronic field, specifically the action functional minimisation (Equation 6). The morphic field of Egyptian warfare, previously tuned to massed infantry and riverine logistics, suddenly finds itself in resonance with a Near Eastern field dominated by highly mobile chariots. Within decades the Egyptian army has not merely adopted chariots; it has made them Egyptian. Pharaoh is depicted in the classic smiting pose while charging in a chariot; the horses are consecrated to Amun; the technology is integrated into the existing ritual order without raising the action function $S(\gamma)$ above tolerable levels.

A similar process occurs with religion. The rise of Amun to pan-Egyptian prominence during the Eighteenth Dynasty does not displace older gods; it overlays them. Amun fuses with Ra to become Amun-Ra, king of the gods. Thebes becomes the new religious capital, yet the ancient sacred sites retain their prestige. Ancient prayers and invocations still refer to Abydos. New temples are built on a scale never previously attempted, but their ground-plans and ritual sequences remain deeply traditional. The morphic field andjusts elastically to accommodate increased wealth and power without losing its fundamental tone.

Akhenaten's reign (1353–1336 BCE) represents the most radical challenge the Pharaonic morphic field ever faced. The solar disk deity Aten is proclaimed sole god; the name of Amun is chiselled from monuments; the capital is moved to a virgin site at Amarna; art abandons millennia-old canons for a fluid, intimate naturalism. For perhaps seventeen years the established morphic fields are deliberately suppressed by continuous coercive forcing. The trajectory (γ) is driven far up the side of the attractor basin defined by the potential term V encoding Ma'at and polytheistic divine kingship (Equation 6).

Yet the moment Akhenaten dies, the restoring force—proportional to the depth of the basin carved by many centuries of polytheistic repetition—is enormous. The resonance score (Equation 7) for the traditional pattern is orders of magnitude higher than for the Amarna experiment. Tutankhamun restores the temples; Horemheb erases the very name of the heretic pharaoh; the Ramesside kings build on a scale that dwarfs even the Eighteenth Dynasty. The field had merely been compressed. Released from pressure, it rebounds with redoubled strength, exactly as Equation 8 predicts for a system returning to the minimum of a deep potential well.

This episode demonstrates a crucial property of morphic fields: they possess powerful inertia. A pattern that has been repeated for centuries or millennia builds up a habitual tendency that is not easily overturned. The second-order term in the resonance kernel (Equation 4) creates smooth grooves along frequent historical paths; Akhenaten's path lay outside those grooves and required continuous energy input to maintain.

Late Period and Ptolemaic Transformation

From the Twenty-Sixth Dynasty onward Egypt comes under increasing foreign influence—Libyan, Nubian, Persian, and finally Ptolemaic Greek. Yet even conquest does not destroy the ancient field. Psammetichus I employs Greek mercenaries but revives Saite art in conscious imitation of the Old Kingdom. The Kushite kings of the Twenty-Fifth Dynasty present themselves as more Egyptian than the Egyptians, building pyramids and restoring temples. Persian kings have themselves depicted on temple walls in pharaonic regalia, offering to Egyptian gods.

With Alexander's conquest of Egypt in 332 BCE the pharaonic morphic field enters its final phase of transformation. It is still incredibly powerful. Alexander is proclaimed the son of Amun Ra after visiting his Oracle in the Siwa Oasis, and is depicted as Osiris in hieroglyphic inscriptions. His successors the Ptolemies adopt the titles, iconography, and ritual obligations of pharaohs. Indeed, they are pharaohs. Cleopatra VII speaks the Egyptian language, is initiated into the mysteries of Isis at Philae, and presents herself to her subjects as a living goddess. Greek philosophy, coinage, and administrative practice are introduced, yet they do not replace the older patterns; they resonate alongside them in a broadened morphic field. The new syncretic deity of the Ptolemies, Serapis, combines aspects of Osiris-Apis with Greek Zeus and Hades, creating a god who can be worshipped in Greek cities and Egyptian temples alike.

The daily temple offerings continue until the edicts of Theodosius in the late fourth century AD. The last known hieroglyphic inscription is dated to 394 AD, more than four thousand years after the first tentative signs were carved on Predynastic labels. Only when two new processes reach critical mass—(1) the administrative replacement of Egyptian by Greek and later Latin and Aramaic, and (2) the accumulation of entirely new morphic fields by Christianity and Islam—does the ancient morphic basin gradually fill and a new attractor landscape emerge.

A Field Still Resonating

The extraordinary persistence and adaptability of Egyptian culture finds an unexpectedly precise mathematical analogue in the nine equations presented in the Annex. The morphic field of Egyptian civilisation appears as a deep attractor basin in three-dimensional cultural time, a valley carved by three millennia of repeated practice. Minor perturbations—foreign invasions, failed inundations, the Amarna heresy—correspond to temporary excursions up the sides of the basin. The gradient of the field, shaped by morphic resonance from the entire past, continually pulls the culture back toward its established habits.

Quantum-like superpositions in the c2 and c3 axes allow multiple cultural possibilities to coexist in latent form such as monotheism during the New Kingdom and Hellenic influence long before Alexander. Entanglement-like correlations across the manifold explain the rapid kingdom-wide adoption of innovations once they cross a threshold of repetition. The mathematics treats resonance as a non-local kernel weighted by past similarity, yielding equations remarkably similar in structure to those governing quantum fields—yet operating on the scale of cultures rather than particles.

Stand today inside the hypostyle hall at Karnak as the evening light strikes the lotus capitals, or read aloud the Negative Confession from the Book of the Dead, and you feel something of the ancient Egyptian morphic field's resonance. The morphic cloud of Pharaonic Egypt did not vanish in 30 BCE, nor even in 394 CE. Traces remain in liturgy of the Copic church, the iconography of the Virgin Mary with the infant Jesus, so similar to the infant Horus in the arms of Isis, and in the continued reverence felt by modern Egyptians for the Nile and its ancient rhythms, now altered by the Aswan Dam.

Nature is habitual creativity. The same processes that enable a salt crystal to form in a supersaturated solution, or a pigeon to reach home across hundreds of miles, enabled a vast and beautiful civilization to maintain its identity through invasion, famine, and religious revolution for over three millennia. The laws of memory are built into the fabric of the world. Ancient Egypt remains one of the clearest demonstrations that the past is not simply gone; under the right conditions of similarity—measured rigorously by the cultural temporal metric—it is still present, tuning the present by morphic resonance to the habits of those who went before.

Mathematical Annex: The Nine Equations of Cultural Morphic Resonance

1. **Cultural Temporal Metric**

   $$ds^2 = g_{ij}\,\Delta c^i\,\Delta c^j$$

   This equation defines the real "distance" between any two moments in Egyptian history: distance is small only when the societies are functionally similar in the same phase (unification vs. unification, pyramid-building vs. pyramid-building, etc.), regardless of how many centuries separate them. It is what allows a Third-Dynasty architect to resonate strongly with a Saite-restoration architect but only weakly with a Mesopotamian ziggurat builder.

2. **Morphic Field Accumulation**

   $$M^i(c^k) = \int K^i{}_j(c^k, c'^k)\,\psi^j(c'^k)\,d^3c'$$

   This is the master equation that builds the collective memory field by summing the final institutionalised patterns (laws, art canons, rituals) of every past successful dynasty, weighted by how similar each past dynasty was to the present moment. It is the mathematical statement of "the past tunes the present."

3. **Gaussian Weight**

   $$G(ds^2) = \exp\left(-\frac{ds^2}{2\sigma^2}\right)$$

   This bell-curve factor ensures that influence from the past falls off smoothly as cultural similarity decreases, so nearly identical historical situations exert a powerful pull while very different ones contribute almost nothing. It is why Old Kingdom pyramid proportions dominate the field but faint echoes of Predynastic palettes still whisper at the edges.

4. **Resonance Kernel**

$$K_{ij}(c^k, c'^k) = G(ds^2)\,\delta_{ij} + \kappa\,\partial_i\partial_j G(ds^2)$$

This kernel combines a direct tug toward identical past patterns with an additional gentle guidance along the smoothed grooves created by thousands of slightly varying but similar histories, producing the characteristic canalisation that makes Egypt keep returning to Ma'at and divine kingship.

5. **Geodesic in Morphic Cloud**

$$\exp_p(v) = \gamma(1), \quad \frac{D\gamma}{d\lambda} = M^i(\gamma)\frac{\delta}{\delta c^i}$$

This equation says that a society naturally evolves along the path where the accumulated morphic field is densest and strongest—just as water flows down the deepest river channel carved by past floods. It explains why unification under Narmer or Mentuhotep II happened so rapidly once the field was ready.

6. **Action Functional**

$$S[\gamma] = \int \left[\frac{1}{2} g_{ij}\dot{\gamma}^i\dot{\gamma}^j - V_{\text{core}}(\gamma) + \kappa\,\delta\psi \cdot M(\gamma)\right] d\tau$$

Nature chooses the historical path that minimises this total "cultural effort" score: paths that stay close to core traditions (Ma'at, divine kingship) and to the average of past successful dynasties cost far less effort than radical departures. It is why foreign rulers from Kushites to Ptolemies found it easiest to rule as pharaohs.

7. **Resonance Score**

$$\mathcal{R} = \int \delta\psi_i(c^2, c^3)\, M^i(c^1, c^2, c^3)\, dc^2\, dc^3$$

This single number predicts how well any proposed future cultural pattern will align with the memory field built during all previous phases; after Akhenaten it gave an overwhelmingly higher score to restored polytheism than to continued Aten-worship, correctly forecasting the explosive backlash.

8. **Equation of Motion**

$$\frac{D^2\gamma^k}{d\tau^2} + \Gamma^k_{ij}\,\dot{\gamma}^i\,\dot{\gamma}^j = -\partial^k V_{\text{core}} + \kappa\,K^k{}_l\,\delta\psi^l$$

This is Newton's second law for entire societies: cultural acceleration is driven by two forces—pull back to the core traditions of Ma'at and kingship plus a corrective tug from the entire resonant past—explaining both the inertia of tradition and the rapid corrections after periods of deviation.

9. **Morphic Field Update**

$$\frac{\partial M^i}{\partial c^1} = \kappa\,\delta^{(3)}(c^k - \gamma^k(\tau))\,\psi^i(\gamma(\tau))$$

Only dynasties that successfully complete all three phases (unification → institutions → religious consolidation) are allowed to stamp their final pattern back onto the field for future generations; failed experiments (most First Intermediate Period micro-kingdoms, the Amarna period) contribute nothing, ensuring the field evolves adaptively yet conservatively.

These nine equations, taken together, transform the poetic idea of morphic resonance into a precise geometric theory that accounts for the astonishing coherence, resilience, and bounded creativity of Pharaonic civilisation across four millennia.

XI

Of Morphic Resonance and Jungian Archetypes

The human experience is a tapestry woven from threads of instinct, imagination, and shared stories that transcend time and culture. Carl Jung, a visionary psychologist, proposed that beneath our individual consciousness lies a collective unconscious, a reservoir of universal patterns called archetypes—the Hero, the Mother, the Trickster, the Shadow—that shape our thoughts, emotions, and behaviors. These archetypes manifest in myths, dreams, and rituals, appearing as the heroic quests of ancient epics or the maternal figures revered across civilizations. Yet, how do these timeless patterns persist, connecting humanity across millennia, and how do they adapt to specific cultural groups? To explore this mystery, we turn to Rupert Sheldrake's theory of morphic resonance, which posits that nature possesses a collective memory, and the mathematical framework of three-dimensional time, which describes a temporal landscape where memory and form intertwine.

This essay reimagines Jung's archetypes as resonant patterns within a human-specific morphic field, a cosmic archive of psychological forms preserved in the three-dimensional temporal manifold. In the mathematics of morphic resonance, we model archetypes as dynamic configurations of humanity's collective memory, accessible universally and, in nuanced variations, linked to specific groups. Quantum-like properties—superposition and entanglement—enable these patterns to connect individuals across time, mirroring the non-local interactions of quantum fields. By weaving together mathematics, biology, physics, and psychology, we uncover a universe where the psyche's deepest patterns move through time, shaping human experience across generations.

Jung's Vision of the Collective Unconscious

Carl Jung's theory of the collective unconscious offers a profound insight into the human psyche. Unlike the personal unconscious, which holds individual memories and experiences, the collective unconscious is a shared reservoir of inherited patterns, universal across humanity. Archetypes, the core elements of this realm, are not specific images but primordial instincts and structures that shape how we perceive and interact with the world. The Hero archetype, for instance, appears in the epic journeys of Gilgamesh, Odysseus, or modern figures like Indiana Jones, embodying courage and transformation. The Mother archetype manifests as nurturing figures like Isis in Egyptian mythology or the Virgin Mary in Christianity, symbolizing care and creation. The Trickster, seen in figures like Loki or Coyote, disrupts norms with cunning and humor, while the Shadow represents the hidden, often darker aspects of the self.

These archetypes are not learned but innate, emerging in dreams, myths, and art across cultures, suggesting a shared psychological heritage. Jung observed their universality in the recurring symbols of diverse societies, yet he also noted their adaptability, taking unique forms in specific cultural contexts—for example, the Trickster as Coyote in Native American lore versus the Fool in European folklore. This dual nature—universal yet particular—raises a question: how are archetypes stored and accessed across generations, and how do they manifest distinctly in groups? To answer, we draw on Sheldrake's morphic resonance, which provides a mechanism for collective memory, and three-dimensional time, which offers a mathematical canvas for modeling this process.

Sheldrake's Morphic Resonance: A Collective Memory

Rupert Sheldrake's theory of morphic resonance proposes that nature possesses a collective memory, stored in species-specific fields called morphic fields. These fields act like invisible blueprints, guiding the development of organisms and behaviors beyond the information encoded in genes. Sheldrake likens genes to a recipe for building materials—sufficient for crafting proteins but inadequate for the full architecture of a complex organism, such as the intricate folding of a bird's wing or the branching of a tree. The morphic field, residing in a species-specific morphic cloud, holds these holistic patterns, connecting individuals across time and space. For example, when blue tits in 20th-century Britain learned to open milk bottles, the behavior spread rapidly, suggesting a shared influence beyond physical communication. Similarly, crystals like insulin form more readily after the process has been repeated many times, as if the morphic field has learned from repetition, reinforcing established patterns.

For Jung's archetypes, morphic resonance offers a framework to envision the collective unconscious as a human-specific morphic field, a repository of psychological patterns that resonate across generations. The Hero archetype, for instance, could be a stable pattern within this field, influencing individuals to dream or create stories of heroic journeys. Group-specific variations, such as culturally distinct Trickster figures, suggest localized fields within the broader human field, shaped by shared histories. To formalize this idea, we need a mathematical structure that supports such a memory, capable of bringing past and present together in a unified physical framework.

Three-Dimensional Time: A Canvas for Reality

The theory of three-dimensional time reimagines the fabric of reality, proposing that time is not a single linear thread but a manifold with three distinct dimensions. The first time dimension, akin to our everyday experience, tracks the progression of moments, like the ticking of a clock. The second dimension represents possibilities, the potential paths a system might take—for example, the myriad ways a psychological impulse might manifest, such as a heroic act or a creative endeavor. The third dimension governs transitions between these possibilities, orchestrating shifts in development, much like a conductor guiding a symphony through its movements. Together, these dimensions form a three-dimensional temporal manifold, a mathematical space where reality unfolds.

As originally developed by the physicist Gunther Kletetschka, this framework unifies General Relativity, which describes gravity as the curvature of four-dimensional spacetime (three spatial, one temporal), with Quantum Field Theory, which models particles as excitations in fields governed by probabilistic wave functions exhibiting superposition (existing in multiple states) and entanglement (correlated states across distances). The theory extends spacetime to a six-dimensional manifold—three fundamental temporal dimensions and three spatial dimensions, with spatial coordinates projected from temporal ones via a projection operator, ensuring causality, with causes preceding effects. In the adapted form of the theory developed in this book, physical and psychological entities, from atoms to archetypes, are dynamic patterns, or worldlines, tracing paths through the temporal manifold. An individual's psychological development, for instance, is a worldline weaving through the three dimensions of time, shaped by personal experiences and collective influences.

The temporal manifold is described using tensors, mathematical arrays that encode interactions across the time dimensions. The metric tensor acts as a ruler, measuring temporal distances between events, much like a map measures spatial distances. This structure supports the morphic field's operation, allowing archetypes to resonate across time, akin to quantum fields' non-local interactions. By integrating quantum principles like superposition and entanglement, our framework provides a robust foundation for modeling the collective unconscious as a field-mediated memory, where archetypes persist and influence the present.

Archetypes as Morphic Patterns in the Temporal Manifold

To model Jung's archetypes, we envision the collective unconscious as a human-specific morphic cloud, a submanifold within the three-dimensional temporal manifold. This cloud stores the psychological patterns of all humans, with archetypes as dominant, stable configurations that resonate strongly due to their recurrence across generations. The morphic field for humanity captures these patterns at each temporal point, constructed by integrating contributions from past individuals' psychological worldlines. The form configuration, a vector analogous to a quantum wave function, encodes the probability distribution of archetypal patterns, such as the Hero's narrative structure or the Mother's nurturing instinct. Superposition allows these patterns to exist in multiple potential forms, while entanglement-like correlations link individuals to ancestral experiences, explaining the universality of archetypes like the Hero across disparate cultures.

For specific human groups, a subset morphic field focuses on patterns unique to them, such as cultural myths or rituals. The resonance process, governed by a kernel, weights the influence of past patterns based on their temporal proximity, ensuring that archetypes like the Mother remain stable over millennia, while group-specific variations, like the Trickster's cultural expressions, adapt to localized histories. The kernel's structure, incorporating a Gaussian function and its derivatives, mirrors the fading of memory with temporal distance and guides the psyche toward archetypal forms through the dimensions of possibility and transition.

The Dance of Resonance: How Archetypes Manifest

The manifestation of archetypes in an individual's psyche is a dynamic process, akin to an embryo developing under genetic and morphic influences. An individual's psychological worldline traces his or her development through the temporal manifold, shaped by personal experiences (encoded as a psychological potential) and resonant influences from the collective unconscious. The morphic field acts like a guiding force, aligning the individual's psyche with archetypal patterns. For example, a dream featuring a heroic figure may reflect resonance with the Hero archetype, collapsing multiple potential expressions into a specific narrative shaped by personal and cultural context.

Quantum-like principles enhance this process. Superposition allows the form configuration to hold multiple archetypal expressions simultaneously, such as the Hero as warrior, savior, or explorer, until a specific context selects one. Entanglement-like correlations connect the individual to the collective unconscious, enabling non-local influences where past psychological patterns shape present experiences, much like quantum particles sharing correlated states across distances. The resonance kernel, with its Gaussian decay, ensures that temporally closer patterns, such as recent cultural myths, exert stronger influence, while ancient archetypes like the Mother retain enduring power due to their stability.

Group-specific archetypes emerge from localized morphic fields, which integrate the psychological worldlines of a community. For instance, the Trickster archetype in Native American cultures, embodied by Coyote, resonates within a group-specific field, reflecting shared stories and values. The mathematical framework predicts that such patterns are reinforced through repeated expression, much like Sheldrake's examples of crystallization or learned behaviors spreading rapidly within a species.

Resolving the Mystery of Archetypal Persistence

Jung's archetypes pose a challenge: how do universal patterns persist across generations, and how do they adapt to specific groups? The morphic resonance model, grounded in the three dimensional time framework, offers an elegant solution. The human morphic field aggregates the psychological history of humanity, with archetypes as stable, resonant patterns amplified by their recurrence. The form configuration's wave function-like nature, supporting superposition and entanglement, allows archetypes to exist as potential forms, collapsing into specific expressions through resonance. Group-specific fields, with narrower temporal parameters, account for cultural variations, such as the distinct Trickster figures in different traditions.

This model yields testable predictions. Cross-cultural studies of dreams or myths may reveal consistent archetypal patterns, such as the Hero's journey, suggesting resonance with a universal morphic field. Rapid adoption of certain narratives in modern media, like superhero stories, could reflect amplified resonance within the collective unconscious, akin to Sheldrake's crystallization phenomena. Group-specific behaviors, such as ritual practices unique to a culture, may strengthen within localized fields, observable through anthropological research.

Philosophical Implications

This synthesis of Jung, Sheldrake, and three dimensional time transforms our understanding of the psyche, bridging psychology, biology, and physics. The collective unconscious becomes a temporal hologram, a dynamic field where archetypes resonate across the dimensions of time, linking past and present in a quantum-like dance. The model resonates with philosophical traditions, such as Plato's forms or Eastern concepts of universal consciousness, yet invites empirical exploration through patterns in dreams, myths, or cultural evolution. It suggests that memory is a fundamental property of the universe, woven into the fabric of time, and that human experience is a symphony of shared patterns, orchestrated by the interplay of individual and collective forces.

By formalizing archetypes as morphic patterns, we glimpse a universe where the psyche is not isolated but interconnected, shaped by a timeless archive of human experience. The mathematics of resonance, guided by the Gaussian kernel and quantum principles, reveals a cosmos where the Hero, the Mother, and the Trickster endure, guiding humanity through the eternal dance of time and memory.

XII

The Eternal Crown: Royalty, Aristocracy, and the Resonant Archetypes of Power

In the grand tapestry of human history, few threads gleam as brightly as those of royalty and aristocracy. From the opulent courts of Vienna's Habsburgs to the sacred rituals of Japan's Yamato dynasty, these exalted lineages have woven narratives of power, continuity, and myth that endure through centuries of revolution and change. Their crowns spark awe, their ceremonies captivate millions, and their legacies anchor societies in an ever-shifting world. Why do the Habsburgs, though stripped of their empire, still stir Europe's imagination with whispers of imperial grandeur? Why does Japan's imperial line, unbroken for millennia, remain the spiritual heartbeat of a modern nation? To unravel this mystery, we turn to Carl Jung's theory of archetypes, the universal patterns of the collective unconscious, and Rupert Sheldrake's morphic resonance, which proposes a collective memory that transcends time. These frameworks reveal royalty and aristocracy as living archetypes, their enduring appeal sustained by a shared psychological heritage that echoes across generations.

This essay explores the concepts of royalty and aristocracy through the lens of Jungian archetypes, using the Habsburg and Yamato dynasties as exemplars to illuminate their remarkable persistence. Jung's archetypes—primordial images like the King, Hero, or Wise One—explain their psychological power, while Sheldrake's morphic resonance offers a mechanism for their cultural endurance, envisioning dynasties as resonant patterns in a collective morphic memory field. Through this synthesis, we can see royalty and aristocracy not as historical relics but as timeless archetypes, enduring through the ages to inform and shape the human experience.

The Archetypal Power of Royalty and Aristocracy

Royalty and aristocracy are more than titles or bloodlines; they are archetypes—universal patterns in the collective unconscious, as Carl Jung proposed, that shape how we think, feel, and act across cultures. The King archetype embodies authority, order, and divine connection, as seen in figures like the Emperor Augustus Caesar, founder of Imperial Rome, or Japan's First Emperor Jimmu Tenno, founder of the Yamato dynasty, who was mythically descended from the sun goddess Amaterasu. The Hero archetype, often a prince or knight, represents courage and transformation, evident in tales of Habsburg emperors defending Christendom or samurai upholding honor. The Wise One archetype, embodied by advisors like Cardinal Richelieu or Japan's Fujiwara regents, channels knowledge and strategy, guiding realms through turmoil. These archetypes resonate because they fulfill deep human needs: the King offers stability in chaos, the Hero inspires action, and the Wise One provides wisdom amid uncertainty.

Jung argued that archetypes are not learned but inherited, emerging in myths, dreams, and rituals worldwide. The King's crown, whether the Habsburg Reichskrone or Yamato's chrysanthemum seal, symbolizes cosmic order, linking mortal life to the divine. Aristocracy, as the King's circle of nobles, extends this archetype into a hierarchy of loyalty and stewardship, from medieval vassals swearing fealty to modern peers gracing ceremonial halls. Their appeal is dual: universal, tapping humanity's shared psyche, yet particular, adapting to cultural contexts like the Habsburgs' Catholic pageantry or Japan's Shinto traditions. This persistence suggests a mechanism beyond history—a collective memory where past patterns shape the present, making dynasties enduring symbols of identity and aspiration.

Sheldrake's Morphic Resonance: A Memory of Majesty

Rupert Sheldrake's theory of morphic resonance offers a compelling explanation for this mechanism. He suggests that nature holds a collective memory in morphic fields, invisible frameworks that transmit patterns of form and behavior across time and space. In biology, these fields guide development beyond genes, as seen in crystals forming more readily after prior formations or behaviors spreading rapidly, like blue tits learning to open milk bottles across 20th-century Britain. Applied to human culture, morphic resonance posits that societal patterns—rituals, hierarchies, myths—grow stronger through repetition, resonating like echoes in a vast temporal hall. For royalty and aristocracy, each coronation, marriage, or noble procession strengthens this field, making future enactments more compelling, more habitual.

In Jungian terms, the collective unconscious is a human-specific morphic field, a reservoir of archetypes like the King or Hero, accessible to all yet expressed uniquely in specific groups. The Habsburgs' imperial ceremonies or Japan's harvest rituals resonate within this field, reinforcing dynastic authority as a cultural touchstone. Sheldrake observes that memories are transmitted through culture, explaining why royal weddings or aristocratic scandals captivate millions: they tap a shared psychological heritage, evoking ancestral attachment. The King archetype resonates in reverence for a monarch's portrait, binding societies to their ancient origins.

The Temporal Stage of Dynastic Resonance

To understand how these archetypes endure, we envision a temporal framework where past and present intertwine. Royalty and aristocracy are not static but dynamic, their stories unfolding as geodesic paths through time, shaped by cultural forces and collective memory. The King archetype, for example, adapts across eras—from warrior-king to ceremonial symbol—while retaining its core as a symbol and source of order. This adaptability suggests a deeper structure, where dynastic patterns are preserved and amplified, much like memories that grow vivid with retelling.

Sheldrake's morphic fields provide this structure, envisioning dynasties as collections of historical events—coronations, battles, marriages—that resonate forward, influencing modern perceptions. The Habsburgs' grandeur, embodied in Vienna's palaces, or the Yamato's rituals, woven into Shinto shrines, act like cultural channels, patterns that persist because they are repeatedly enacted. These fields are universal for humanity, capturing archetypes like the King, yet localized for groups, adapting the archetype to cultural nuances, such as Catholic Habsburgs versus the Shinto Yamato dynasty. The resonance process, where past glories echo into the present, explains why dynastic symbols—crowns, crests, ceremonies—evoke such fascination, connecting individuals to a shared ancestral narrative.

The Habsburgs: Archetypes of Imperial Ambition

The Habsburg dynasty, Western Europe's architects of empire, embodies the archetypal power of royalty and aristocracy. Originating around 1020 CE with Radbot's castle in what is now Switzerland, their rise began with Rudolf I's election as German king in 1273, securing Austria as their heartland. Their genius lay in strategic marriages: Maximilian I's 1477 union with Mary of Burgundy brought the Netherlands, while Philip the Handsome's marriage to Joanna the Mad fused Habsburg blood with Spain's, birthing Charles V's vast empire, from Vienna to the Americas. Their motto, "Austriae est imperare orbi universo" (Austria is destined to rule the world), captured the King archetype's ambition, blending divine sanction with earthly dominion.

In Jungian terms, the Habsburgs personified the King as unifier, their coronations and imperial diets resonating as rituals of cosmic order. The Hero archetype shone in Charles V, who defeated Ottoman expansionism, while the Wise One emerged in Maria Theresa, whose 1740 reforms preserved the empire against Prussian threats. Their aristocratic allies—noble houses like the Fugger or Medici—extended these archetypes, acting as stewards and patrons, their wealth and art amplifying the dynasty's luster. Yet the chaos of World War I fatally undermined their rule, and Emperor Franz Joseph's death in 1916, in the midst of the war, was followed by the dissolution of the empire at its end in 1918.

Their appeal endures in modern Europe, where Schönbrunn's gardens and the Hofburg's relics evoke a lost golden age of diversity within unity, oddly resonating with the European Union's ethos of harmony in diversity. Museums and cultural festivals retell Habsburg stories, sustaining the King and Hero archetypes as symbols of a shared past. This persistence reflects a collective memory where persistent rituals, like the Order of the Golden Fleece, still reinforce dynastic patterns, making the Habsburg legacy a cultural balm against modernity's fragmentation.

The Yamato Dynasty: Archetypes of Sacred Continuity

Japan's Imperial House, the Yamato dynasty, offers a contrasting archetype of enduring resonance. Mythically founded in 660 BCE by Jimmu, descendant of the sun goddess Amaterasu, it is the world's oldest continuous monarchy. Unlike the Habsburgs' expansionism, Yamato strength lies in spiritual continuity, rooted in Shinto's animistic worldview. The Emperor as high priest holds the highest ceremonial position in the Shinto religion. He presided over the sacred Niiname-sai rice harvest ceremony, as well as the ceremony in February, which is designed to assure a bountiful rice crop. Thes3 and other rituals resonate profoundly with Japan's historical roots. In

Japan the King archetype functions less as conqueror and more as sacred symbol and priest-king. This role was formalized for the contemporary world in the 1947 Constitution's Article 1, which designated the emperor as the "symbol of the state and unity of the people."

The Hero archetype emerges in the figure of Prince Shotoku. His 7th-century CE governance reforms and promotion of Buddhist culture were so successful that he is considered an incarnation of the bodhisattva of compassion. The archetype of the Wise One can be seen in the Fujiwara regents, who created the refined courtly culture of the Heian period. Emperor Hirohito's formal renunciation of divinity in 1945, a pivotal shift, adapted the King archetype to democratic Japan, a development amplified by Emperor Akihito's empathetic broadcasts during the 2011 tsunami. Today, Naruhito's Reiwa era (meaning "beautiful harmony") sustains this resonance, with polls showing public support ranging between 70 and 80 percent. Imperial births or weddings are still celebrated as national events.

The Resonance of Archetypes: Why Dynasties Endure

The enduring appeal of royalty and aristocracy lies in their archetypal power, amplified by morphic resonance. The King archetype, universal across cultures, provides a sense of order and divine connection, while the Hero and Wise One inspire action and wisdom. These patterns resonate strongly because they are reinforced through repetition—each coronation, each noble deed, strengthens their hold on the collective psyche. The Habsburgs' imperial splendor, etched in Europe's memory through palaces and art, or the Yamato's sacred rituals, woven into Shinto rituals, act as cultural touchstones, their resonance adapting to new eras while retaining their core meaning.

Group-specific variations explain cultural differences. The Habsburgs' Catholic pageantry shaped the King as a warrior-unifier, resonating in Europe's competitive courts, while the Yamato's Shinto roots cast the King as a spiritual guardian, fostering Japan's introspective unity. This adaptability ensures dynasties remain relevant, whether as modern celebrities (e.g., Britain's Windsors) or national symbols (e.g., Japan's emperors). Their resonance is evident in the public's fascination with them. Global interest in royal weddings or the enduring allure of aristocratic estates reflects a shared psychological heritage, a collective unconscious in which archetypes live.

Challenges persist—Habsburg inbreeding or Japan's succession debates highlight vulnerabilities—yet their appeal endures. Modern media, from tabloids to Instagram, humanize royals, blending the mythic with the mundane, while cultural institutions like museums or festivals sustain aristocratic legacies. The morphic field, enriched by centuries of ritual, ensures that dynastic patterns remain vivid, their archetypes resonating as society's anchors.

Philosophical Implications

Royalty and aristocracy, as Jungian archetypes, bridge psychology and culture, their resonance a testament to humanity's need for mythic order. They echo Plato's forms, eternal ideals shaping reality, or Eastern mandalas, symbols of cosmic harmony. Their persistence challenges egalitarian ideals, yet their adaptability—Habsburgs as cultural icons, the Yamato as national unifiers—shows their flexibility. Philosophically, they align with Edmund Burke's vision of society as a partnership across generations, their rituals a living link to distant ancestors in the modern age. In an era of artificial intelligence and transience, the appeal of crowns surges as a call to reconnect with shared myths. The King, the Hero, and the Wise One, whether in Vienna's palaces or Kyoto's shrines, remain beacons, their archetypes arching through time to continue weaving humanity's story.

XIII

The Remarkable Endurance of Armenia

The Armenian people have existed as a recognisably continuous ethnic, linguistic, and religious entity for at least three thousand years, and possibly for much longer, as recent archeological discoveries in Armenia suggest. From the first mention of the "land of Armina" in the trilingual inscription of Darius I at Behistun in 521 BCE, through the Urartian highland civilization that immediately preceded them, the adoption of Christianity in 301 CE, the invention of the Mesropian alphabet in 406 CE, the medieval kingdoms of Bagratid Ani and Cilician Het'um, the centuries of Persian and Ottoman rule without sovereignty, the genocide of 1915–1923, the Soviet period, and the rebirth of an independent republic in 1991, the same language, the same melodic modes of the sharakan hymns, the same sense of covenant with the mountain called Masis (Ararat), and the same stubborn conviction of being a people called the "Hayk" have endured. No conqueror—Persian, Roman, Byzantine, Arab, Seljuk, Mongol, Ottoman, Russian, or Soviet—has succeeded in dissolving this identity. The same motifs recur across millennia: the khatchkar with its radiating symbol of eternity, the vine-scroll that symbolizes the resurrection of the dead and eternal life, the liturgy sung in Grabar, the classical tongue that is to modern Armenian what Latin is to Italian—yet still understood in every village church. Improbably, remarkably, Armenian civilization endures.

Conventional explanations invoke mountain isolation, the protective role of the Church, the diaspora merchant networks, and the emotional charge of collective memory after the Genocide. These are real, yet they remain local, energetic mechanisms: priest to deacon, merchant to merchant, survivor to child. Something more pervasive, something non-local and non-energetic, has made certain thoughts, gestures, melodies, and practices feel self-evidently "Armenian" to people separated by two thousand years, and often by continents. The hypothesis of morphic resonance, now expressed with full mathematical rigour through a three-dimensional cultural time framework, offers the explanatory bridge (see the Mathematical Annex).

Cultural evolution does not march along the single arrow of clock time. It moves within a three-dimensional temporal manifold whose coordinates represent distinct dimensions of Armenian cultural organisation:

- c^1: unification and self-identification as Hayk
- c^2: institutional and liturgical elaboration
- c^3: symbolic and territorial maturation (mountain, diaspora, memory)

The functional distance between any two cultural states is given not by calendar years but by the cultural temporal metric (Equation 1). Only states that are genuinely homologous across these three dimensions have a small ds^2 and can therefore resonate strongly. The Gaussian weight (Equation 3) ensures that morphic influence decays smoothly with increasing cultural dissimilarity, yet for Armenians the width parameter σ is unusually small; the pull from the fifth century CE remains almost undiminished today.

The Emergence of the Field: Urartu to the Artaxiad Kingdom

The Armenian Highland was already a distinct cultural province when the kingdom of Urartu flourished between the ninth and sixth centuries BCE. Fortresses ringed Lake Van, irrigation canals watered the plain of Ararat, and cuneiform inscriptions invoked the god Ḫaldi. When Urartu collapsed under Median and Scythian pressure, the material culture did not vanish; it was inherited by the satrapy of Armina under the Achaemenids and then by the Orontid (Yervanduni) dynasty. The same volcanic stone, the same terraced agriculture, the same highland defensive habits persisted. The morphic field was already taking shape. As local chiefs buried their dead with the same bronze belts and carved the same lion motifs, the cumulative past strengthened the field (Equation 2). The resonance kernel (Equation 4) began to carve the first deep grooves: mountain redoubt, covenant with sacred landscape, and a sense of chosen peoplehood that would later be transferred without rupture to Christian Armenia.

Tigran the Great (95–55 BCE) briefly created an empire stretching from the Caspian to the Mediterranean. For a moment the c^1 coordinate of political unification reached a value unmatched until the twentieth century. The societal trajectory shot along the geodesic defined by Equation 5, driven by a morphic field vector that already pointed strongly toward highland-centred kingship. After Tigran's defeat by Rome, the Artaxiad kingdom shrank, but the field did not contract; it simply deepened the attractor basin around the mountain core. The Armenian language, already fully formed as an independent branch of Indo-European, became the vehicle of royal inscriptions and oral epic.

Crystallization through Conversion: The Fourth and Fifth Centuries CE

The adoption of Christianity in 301 CE by King Trdat IV and Gregory the Illuminator was the single most powerful reinforcement the Armenian morphic field has ever received. The new faith was not grafted onto a cultural vacuum; it was tuned to the existing sense of covenant. Pagan temples were converted into churches on the same sacred sites—fire replaced by incense, Ḫaldi by the Cross, yet the landscape itself remained the axis mundi. The resonance score (Equation 7) for Christianity was extraordinarily high because it resonated with the pre-existing c^3 dimension of chosenness and sacred geography.

Then, in 406 CE, Mesrop Mashtots created the Armenian alphabet. The invention was not a technical exercise; it was a morphic event of the first order. Thirty-eight letters were shaped to capture every phoneme of the spoken language with perfect fidelity. From that moment the field acquired a voice that could be fixed in writing yet still felt like speech. Manuscripts proliferated; the Bible was translated within a decade; histories were written that consciously placed the Armenians in sacred history from Hayk to Christ. Every completed codex strengthened the field (Equation 9). The action functional (Equation 6) plummeted whenever a trajectory stayed close to this new liturgical-linguistic core. To be Armenian now meant to speak, pray, and write in a way that resonated with the fifth century across every subsequent century.

Medieval Kingdoms and the Deepening of the Groove

The Bagratid kingdom (885–1045 CE) saw the city of Ani rise to a hundred thousand souls and a thousand churches. Architecture reached a pitch of refinement never surpassed: the cathedral of Ani, the church of Aght'amar with its reliefs of David and Goliath, the gavits and khatchkars whose interlaced knotwork seems to grow organically from the stone. The resonance kernel's second-order term (Equation 4) carved an extraordinarily smooth channel along the path of mountain monarchy plus apostolic church. When the Seljuks conquered Ani in 1064, the political structure collapsed, but the field simply decentralised. Monks carried manuscripts to new centres; the pattern of refuge-and-continuation became part of the geodesic itself.

In Cilician Armenia (1080–1375) the field demonstrated its elasticity. Armenian barons ruled a Mediterranean coast, married Frankish princesses, negotiated with Crusader kings, and minted coins bearing both cross and crescent. Yet the Catholicosate remained Armenian, the liturgy was sung in Grabar, the nobility traced their descent from Hayk and the Bagratids. The morphic field stretched across languages and seas without breaking; the same resonance score rewarded fidelity to the mountain-liturgical core even in exile.

The Long Survival without Sovereignty (1375–1918)

For more than five centuries Armenians lived under Ottoman, Persian, or Russian rule without a state. Villages paid the harac tax, kept their priests, and carved khatchkars in secret. In the great cities—Constantinople, Smyrna, Tiflis, Julfa—Armenian merchants formed tight communities around their churches. The Mekhitarist fathers in Venice and Vienna printed the first Armenian Bible in movable type, compiled dictionaries, and sent books by caravan across half Asia. The field was no longer tied to territory; it had become a distributed network sustained by manuscript, liturgy, and endogamy. The equation of motion (Equation 8) continued to exert a restoring force proportional to any deviation from the deep attractor of language-church-mountain-memory.

The Genocide and the Supreme Test of the Field

Between 1915 and 1923 the Young Turk government attempted the total extirpation of the Armenian people from their historic homeland. A million and a half perished; ancient communities from Van to Musa Dagh were erased. By every material measure the field should have been shattered. Yet the very attempt at annihilation deepened the groove of remembrance. The survivors who reached Beirut, Marseille, Boston, or Fresno carried with them an intensified determination to remain Armenian. Within months of arriving in the desert camps of Deir ez-Zor, refugees were founding schools. In 1920s Beirut the first Armenian political parties and newspapers appeared. The Badarak was sung in rented halls; the Mesropian letters were taught to children born in exile. The morphic field accepted the terrible compression and rebounded with redoubled strength, exactly as Equation 8 predicts for a system released from extreme coercive forcing. The resonance score for cultural survival became infinite; assimilation scored negative infinity.

Soviet Armenia and the Hidden Continuity

In the east, historic Eastern Armenia became a Soviet republic in 1920. For seventy years the red flag flew over Yerevan, yet the field adapted with astonishing subtlety. Armenian was an official literary language of the USSR; the Matenadaran preserved thousands of medieval manuscripts; poets wrote in Soviet forms but in Armenian metres; underground priests kept the old calendar. The earthquake of 1988 and the Karabakh movement of 1988–1991 revealed that the field had remained intact beneath the ideological crust. When the Soviet Union dissolved, independence felt less like invention and more like recognition.

Artsakh and the New Diaspora

The Republic of Artsakh (1991–2023) represented a brief but complete maturation in all three cultural coordinates: its own army, anthem, schools, parliament, and khatchkars on every hill. Even its tragic loss in 2023 has already stamped a final ψ vector into the field (Equation 9). The hundred thousand refugees who fled to Armenia in September 2023 are already opening Saturday schools in Yerevan suburbs and building new churches. The groove has simply widened to include another cycle of mountain refuge and diaspora renewal.

A Field Still Resonating

Stand today inside the cathedral of Etchmiadzin as the deacon raises the Gospel in the ancient rite, or watch a master mason in Gyumri carve a new khatchkar whose knotwork is indistinguishable from one carved in 1291, or listen to a child in Glendale recite the fifth-century hymn "Aravot louso" in perfect Grabar intonation, and you feel the Armenian morphic field in full strength. The cloud did not vanish in 1375, nor in 1915, nor in 2023. Traces remain in the unbroken apostolic succession, in the continued use of the Mesropian letters, in the reverence felt by every Armenian for Masis even when it is visible only in memory.

The mathematics in the Annex treats resonance as a non-local kernel weighted by past similarity, yielding equations remarkably similar in structure to those governing quantum fields—yet operating on the scale of peoples rather than particles. Nature is habitual creativity. The same processes that enable a termite mound to maintain its form, or a flock of starlings to wheel as one organism, enabled a small highland nation to maintain its identity through empire, genocide, and exile for over three thousand years. The laws of memory are woven into the fabric of reality. Armenia remains one of the clearest demonstrations that the past is not simply gone; under the right conditions of similarity—measured rigorously by the cultural temporal metric—it is still present, tuning the present by morphic resonance to the habits of those who went before.

Mathematical Annex: The Nine Equations of Armenian Cultural Resonance

1. **Cultural Temporal Metric**

   $$ds^2 = g_{ij}\, \Delta c^i\, \Delta c^j$$

   This equation defines how "far apart" two moments in Armenian history really are. Instead of measuring distance in years, it measures how much the sense of unity, liturgy, and sacred geography have changed; for Armenians this cultural distance has stayed remarkably small for 2,500 years, which is why the past still feels close.

2. **Morphic Field Accumulation**

   $$M^i(c^k) = \int K^i{}_j(c^k, c'^k)\, \psi^j(c'^k)\, d^3c'$$

   The morphic field is the accumulated influence of everything Armenians have ever successfully been and done. This equation adds up all past periods (Urartu, Bagratid Ani, Cilician kingdom, diaspora communities, etc.) and lets the strongest, most similar ones shape what Armenians feel is "natural" today.

3. **Gaussian Weight**

   $$G(ds^2) = \exp\left(-\frac{ds^2}{2\sigma^2}\right)$$

   Past events only influence the present strongly if they are culturally similar to the present. For Armenians the bell curve is extremely narrow (small σ), so a fifth-century monastery service still feels more relevant than a nineteenth-century European lifestyle ever could.

4. **Resonance Kernel**

   $$K_{ij}(c^k, c'^k) = G(ds^2)\, \delta_{ij} + \kappa\, \partial_i \partial_j G(ds^2)$$

   This is the rule that pulls Armenian culture toward patterns that have worked many times before. The extra term smooths the path along the most-traveled routes (mountain refuge → church continuity → diaspora renewal), making those routes feel easier and more "correct" than any other option.

5. **Geodesic in Morphic Cloud**

$$\exp_p(v) = \gamma(1), \qquad \frac{D\gamma}{d\lambda} = M^i(\gamma)\frac{\delta}{\delta c^i}$$

Armenian history does not wander randomly; it follows the single shortest, most-reinforced path through cultural space. The morphic field acts like gravity, guiding society along the trajectory that past Armenians have already made deepest.

6. **Action Functional**

$$S[\gamma] = \int \left[\frac{1}{2} g_{ij}\dot{\gamma}^i\dot{\gamma}^j - V_{\text{core}}(\gamma) + \kappa\,\delta\psi\cdot M(\gamma)\right] d\tau$$

Any possible future for Armenians has an "energy cost." Paths that stay close to the ancient core of language, church, and mountain memory have the lowest cost, so they are the ones that actually happen with the least resistance.

7. **Resonance Score**

$$\mathcal{R} = \int \delta\psi_i(c^2, c^3)\, M^i(c^1, c^2, c^3)\, dc^2\, dc^3$$

This number predicts how viable a proposed future feels to Armenians. After the Genocide, "keep being Armenian at all costs" scored infinitely high while "assimilate and disappear" scored infinitely low, which is exactly what happened in real life.

8. **Equation of Motion**

$$\frac{D^2\gamma^k}{d\tau^2} + \Gamma^k_{ij}\dot{\gamma}^i\dot{\gamma}^j = -\partial^k V_{\text{core}} + \kappa\, K^k_{\ l}\,\delta\psi^l$$

Whenever conquerors or ideologies push Armenian culture away from its ancient core, this equation creates a restoring force—like a spring—that pulls it back. That is why foreign rule, atheism, and even genocide ultimately failed to erase Armenian identity.

9. **Morphic Field Update**

$$\frac{\partial M^i}{\partial c^1} = \kappa\, \delta^{(3)}(c^k - \gamma^k(\tau))\, \psi^i(\gamma(\tau))$$

Only epochs that Armenians fully lived through and completed (such as the Cilician kingdom or the Artsakh republic) get permanently etched into the collective memory field. Failed attempts at assimilation or cultural erasure leave no lasting trace and do not weaken the field.

XIV

The Morphic Cloud of the Roman Empire

We are accustomed to thinking of empires as structures composed of people, laws, roads, armies, and taxes. Yet there is another way of seeing them, less familiar to the orthodox historian and yet, perhaps, far closer to the lived reality of those who built and inhabited them. An empire is also a form, a persisting pattern that organizes the behavior of millions across centuries. Such forms are not imposed solely by genes or by conscious planning; they are sustained by the faculty of memory inherent in nature herself—a memory that acts across time by virtue of morphic resonance.

Just as a foxglove "remembers" how to make its spotted trumpet, or a termite colony "remembers" the exact architecture of its mound, so too did the Roman Empire come to remember how to be Rome. There arose, slowly at first and then with gathering momentum, a morphic field of Romanitas—an invisible yet potent cloud of influence that shaped the thoughts, habits, institutions, and even the very gait of men who never read a line of Livy or set foot inside the Curia. This field did not hover somewhere above the seven hills like a Platonic idea. It was immanent in the repeated rituals, the reiterated oaths, the daily wearing of the toga, the annual cycle of ludi (games) and triumphs, and the habitual obedience of legionaries to the eagle standards. Each time a Roman did what Romans do, he reinforced the field; and the field, in turn, made it ever easier for the next Roman to do the same.

Let us trace, then, the birth and growth of this great morphic cloud from the legendary founding of Rome in 753 BC to the radical restructuring under Diocletian in AD 284–305—an empire that by then had learned, through sheer repetition across twelve centuries, how to be itself. In the beginning there was no cloud—only a handful of mud huts on the Palatine Hill and a few hundred Latins, Sabines, and Etruscans trying to agree on how to live together. Their morphic field's strength was vanishingly small. Yet even then the first faint resonances were laid down.

Romulus kills Remus and opens the "asylum" on the Capitoline, a sacred refuge for any who might want to join the new city. The kidnapping of the Sabine women introduces the idea of exogamy, or the acquiring of brides from neighboring communities, leading to rapid population growth. The kings—Romulus, Numa, Tullus, Ancus, and the Tarquins—perform the same gestures again and again: the tracing of the pomerium with a bronze plough along the borders of a new settlement, the dedication of the spoils of war to Jupiter and Mars, and the consultation of auguries by observing the flight of birds. Each repetition is a minute deposition into the nascent morphic field. The temporal metric is still almost flat (eq. 1); the "distance" in form-space between one king's reign and the next is tiny. Resonance is therefore strong even though the absolute number of past instances is small.

By the time the last Tarquin is expelled in 509 BC, a distinct Roman pattern has crystallised: a city ruled neither by a sole king nor by the chaotic assembly of clans, but by two annually elected magistrates flanked by a senate of "patres conscripti" ("conscripted fathers" — a term signifying wise elders). This peculiar arrangement could perfectly well have been otherwise—many Italian cities remained monarchies or tyrannies—but once the pattern is repeated a dozen times it becomes the natural way to be Roman. Young men growing up after 509 do not need to be told that kings are hateful; the morphic field already pulls them toward the republican solution with a force proportional to the accumulated past.

With the expulsion of the kings the field begins to grow exponentially. Every year two consuls are elected; every five years censors oversee morals and the roll of registered citizens; every decade or so Rome fights the same kind of war against neighbouring cities and incorporates them in the same way—half are made citizens without vote, half receive Latin rights, allies are bound by treaties of strikingly similar wording. The repetition is relentless. Consider the triumph. A victorious general, his face painted red in imitation of Jupiter, rides through the city in a four-horse chariot while his soldiers sing ribald verses. This is not mere pageantry; it is a ritual deposition into the morphic cloud. By the time the Samnite Wars ended in 290 BC, the triumph has been celebrated 100 times. By the end of the Second Punic War in 201 BC, that figure had risen to two hundred. Scipio Africanus did not invent his own behaviour; he was pulled along the geodesic path blazed in the three dimensional time manifold by all previous triumphators (eq. 5).

The same is true of law. The oldest known written compilation of Roman law, the Twelve Tables (450 BC), are succeeded by an unending stream of praetorian edicts, senatus consulta, and plebiscites, yet the style remains unmistakably Roman: terse, pragmatic, and obsessed with both procedure and principle. A young quaestor in 180 BC does not need to read all previous edicts; the field guides his hand toward the correct formulae because the resonance score is maximized when he drafts in the accustomed way (eq. 7).

By the time of the Gracchi the field is so strong that attempts at radical innovation provoke violent recoil. Tiberius Gracchus's land law is not merely opposed on grounds of interest; it feels wrong, a deviation from the ancestral path, known as the "mos maiorum." The Roman aristocracy beats him to death with benches and clubs because the morphic field itself, speaking through their outrage, demanded it. Paradoxically the very strength of the republican field now begins to tear it apart. Success has made Rome too large for the old habits to contain. The Italian allies demand citizenship; the publicani (state contractors) grow fabulously rich; armies are no longer citizen militias but professional forces loyal to their generals. Marius, Sulla, Pompey, Caesar—each in his own way tries to stretch the republican form beyond its elastic limit.

Yet even the most revolutionary actors remain inescapably within the morphic cloud. Sulla does not abolish the republic; he "restores" it by doubling the senate and codifying magistracies. Caesar accepts a dictatorship but still attends the Lupercalia in the old way, still refuses a royal crown in traditionally correct, Republican manner. Even Caesar's assassins on the Ides of March believe they are acting in the spirit of Junius Brutus, who had founded the republic six centuries earlier after expelling the last king of Rome, Tarquinius the Proud. No one can quite imagine a form of power that is not somehow derived from the combination of consulship, tribunate, triumph, and dictatorship. The field has become hypertrophic and over-determined. Too many past instances pull in too many directions. Civil war is what happens when a morphic field enters a state of destructive interference and the action functional no longer possesses a single deep minimum (eq. 6).

Octavian—Augustus—achieves something that looks, from the outside, like a coup d'état but feels, to those inside the field, like the natural order reasserting itself. He does not abolish the republic; he fulfills it. All the old forms are preserved—consuls, senate, comitia, triumphs—but they are nested inside a new monarchical attractor that was always latent within the cloud.

Augustus's genius was to move just far enough from the republican fixed point to escape destructive interference, yet not so far that the Gaussian weight with the entire past became negligible (Equation 3). The new principate settles into a deep basin of attraction governed by the accumulated morphic field (Equation 2). The Julio-Claudian and Flavian emperors reinforce this new pattern with obsessive repetition. The same phenomena are repeated: ancient acclamations (imperator, pater patriae, pontifex maximus), the refusal of certain honors followed by their reluctant acceptance, and the distribution of largesse to the praetorian guard. Even mad Emperors such as Caligula and Nero cannot escape the morphic field of Romanitas; they merely explore its pathological extrema before the equation of motion snaps the system back toward the stable attractor (Equation 8).

Under the Antonine emperors the Romanitas field reaches its maximum coherence and extent. A Greek-speaking notable in Antioch or a Celtic aristocrat in Lugdunum wakes up thinking in Roman categories. He builds a villa with atrium and triclinium, educates his sons in rhetoric and law, and offers sacrifice to the imperial numen alongside the local gods. The resonance is now so strong that even at the periphery the correct forms arise almost spontaneously. The Severan period still belongs to this high-imperial basin. Septimius Severus, a Punic-speaking African, founds a dynasty that feels perfectly Roman because every gesture—dynastic succession through adoption or biology, the building of monumental arches, and the deification of empresses—has been rehearsed innumerable times before.

Then, abruptly, the accumulated weight of the past becomes a mortal danger. The Antonine pandemic, endless frontier wars, monetary debasement, and the inability of the principate to produce a reliable succession shatter the old basin of attraction. Between 235 and 284 at least twenty-six emperors are acknowledged in Rome, and dozens more reign briefly in one part of the empire or another. This is not mere political chaos; it is a crisis of the morphic field itself. The field is still there—stronger than ever in sheer accumulated mass (Equation 2), but the resonance kernel now contains dozens of shallow local minima (Equation 4). Gallienus tries to rule as a Platonic philosopher-king; Aurelian restores the old Severan model and proclaims himself restitutor orbis; Probus and Carus attempt a return to the soldier-emperors of the third century. None of the patterns stick for long.

Diocletian's solution is radical yet deeply conservative. He does not try to return to the principate; that basin is shattered beyond repair. Instead he allows the field to settle into an entirely new configuration—the Dominate—that had been latent within Roman memory from the very beginning. Look closely and you will see that Diocletian's innovations are not innovations at all. The adoration of the imperial purple, the diadem, the elaborate court ceremonial—all are borrowed from the hellenistic east and, further back, from Persia and Egypt. But Rome had encountered these forms before: in the regal period, in the triumph, and in the quasi-divine honors voted to Caesar and his successor Augustus. Diocletian simply reactivates very ancient layers of the morphic cloud.

The tetrarchy itself—two Augusti, two Caesars, in a system of collegial rule tempered by hierarchy—is a brilliant exploitation of the update rule (Equation 9): by having four rulers perform almost identical rituals simultaneously, the field receives four reinforcing depositions at once. Even the persecution of the Christians served the field. For the first time in generations the empire had a single, shared ritual of refusal. Those who sacrificed reinforced the ancestral field with particular intensity. By 305, when Diocletian and Maximian abdicated together at the same moment—an utterly unprecedented act that nevertheless succeeded—the Dominate had become the new natural way to be Roman. The geodesic now pointed unerringly toward divine monarchy, the sacred palace, and bureaucratic hierarchy, foreshadowing the birth of Byzantium (see Chapter Fifteen).

From the plough of Romulus to the abdication at Nicomedia, the Roman Empire was never merely a political structure. It was a cloud of habits—habits ingrained in millions of minds and bodies through sheer repetition, composing a morphic structure that grew stronger with every triumph, every sacrifice, and every recitation of the laws. The equations in the annex are simply a formal way of describing what every Roman felt in his bones: that the past was not dead, but present as a shaping influence; that to act Roman was to be pulled along paths trodden by countless ancestors; and that when those paths finally shattered, new paths could be found by reaching deeper into the same resonant cloud of memory.

The empire would change again—Constantine would graft Christianity onto the field, transforming Romanitas, Theodosius would prune away the old gods, Justinian would attempt a vast reconquest of the West—but the cloud itself endured. Even today, when a judge dons a robe, or a soldier salutes an eagle, or a citizen speaks of the "republic" (res publica), faint echoes still reach us across the centuries. The morphic field of Rome has not dissipated;

it has only grown subtler, more diffuse, waiting for new forms to resonate with its ancient habit.

And that, perhaps, is the deepest lesson Rome teaches: civilizations are not built by heroes or laws alone. They are remembered into being.

Mathematical Annex

The Morphic Resonance of Romanitas

1. **Temporal Metric**

   $$ds^2 = g_{ij}\, \Delta t^i\, \Delta t^j$$

   Measures morphic distance. Crucial in 27 BC: Augustus's restoration of the Republic feels almost identical in form to the consulate of 509 BC, making ds^2 tiny and resonance enormous despite 482 calendar years.

2. **Morphic Field Accumulation**

   $$M^i(t^k) = \int K^i{}_j(t^k, t'^k)\, \phi^j(t'^k)\, d^3t'$$

   By AD 180 the integral contains ~370 consul-pairs, ~220 triumphs, and thousands of standardised treaties. This is why Marcus Aurelius on the Danube instinctively thinks like a Roman gentleman: the entire past presses upon him through M^i.

3. **Gaussian Weight**

   $$G(ds^2) = \exp\left(-\frac{ds^2}{2\sigma^2}\right)$$

   Sulla's dictatorship (82–79 BC) resonates far more strongly with Caesar's (49–44 BC) than with the regal period five centuries earlier; morphic distance to the kings is large.

4. **Resonance Kernel**

   $$K_{ij}(t^k, t'^k) = G(ds^2)\, \delta_{ij} + \kappa\, \partial_i \partial_j G(ds^2)$$

   The ∂∂G term creates directional pull. Each emperor who accepts "dominus" or the diadem stretches the form a little further; by the 290s the kernel has carved a deep valley toward the Dominate.

5. **Geodesic in the Morphic Cloud**

$$\exp_p(v) = \gamma(1), \quad \frac{D\gamma}{d\lambda} = M^i(\gamma)\frac{\partial}{\partial t^i}$$

The path of least resistance. Republic → Civil Wars → Principate → Dominate is the single longest geodesic in the cloud, followed with astonishing straightness from 509 BC to 300 AD and beyond.

6. **Action Functional for Institutional Growth**

$$S[\gamma] = \int \left[\frac{1}{2} g_{ij}\dot{\gamma}^i\dot{\gamma}^j - V_{\text{gene}}(\gamma) + \kappa\,\delta\phi \cdot M(\gamma)\right] d\tau$$

V_gene = material constraints. The Gracchan reforms minimised S too violently against V_gene and were expelled; Augustus found a path satisfying both terms.

7. **Resonance Score**

$$\mathcal{R} = \int_{t^2,t^3} \delta\phi_i(t^2,t^3)\, M^i(t^1,t^2,t^3)\, dt^2 dt^3$$

When Antony placed the diadem on Caesar at the Lupercalia (44 BC) the resonance score plummeted; the gesture lay outside accumulated M^i. The crowd's silence and Caesar's refusal were the field rejecting a low-scoring trial.

8. **Equation of Motion**

$$\frac{D^2\gamma^k}{d\tau^2} + \Gamma^k_{ij}\dot{\gamma}^i\dot{\gamma}^j = -\partial^k V_{\text{gene}} + \kappa\, K^k{}_l\, \delta\phi^l$$

In the third-century crisis the morphic term fragmented while ∂V_gene (plague, invasions, debasement) grew violent, producing fifty years of chaotic acceleration. Diocletian restored coherent κK.

9. **Morphic Field Update**

   $$\frac{\partial M^i}{\partial t^1} = \kappa\, \delta^{(3)}(t^k - \gamma^k(\tau))\, \phi^i(\gamma(\tau))$$

   The simultaneous abdication of Diocletian and Maximian on 1 May 305 constituted an extraordinarily strong Dirac-δ deposition—two identical realisations at once—briefly making peaceful tetrarchic succession the new natural way to be Roman.

Thus the mathematics is not ornament; it is the formal language in which the living past of Rome spoke to its present, and still speaks faintly to ours.

XV

The Morphogenesis of Civilizations: the Transition from Rome to Byzantium

There is a presence in nature that is older than books, older than writing itself. Whenever something has happened once, it is, in some sense, more likely to happen again. A crystal forms more easily because crystals of its kind have formed before; a bird sings its ancestral song more readily because that song has been sung a million times in forests long vanished; a human mother, without being taught, folds her arms around her newborn in the same way mothers have done for countless generations. These are not coincidences. They are traces of an invisible memory that inheres in nature herself. We call this memory morphic resonance: the influence of like upon like across space and time, transmitted neither by genes nor by material traces, but by a morphic field—a field that is within and around each kind of being, a field that grows stronger with repetition and reaches backward and forward through time. Seen in this way, the past does not merely cause the present; it is ever present, guiding and influencing whatever comes next.

When we turn our gaze to the great organisms we name civilizations, we find the same process at work, only vaster in scale and slower in rhythm. Civilizations are not heaps of customs and artefacts. They are living fields of memory, extended in space and stretched across centuries. Their laws, their languages, their cathedrals, their ways of crowning kings—all these are shaped not only by the pressures of the moment but by the accumulated past of their own kind. And when one civilization gives birth to another, as Rome gave birth to Byzantium, we witness a speciation event in the realm of collective mind and form, a process directed by the same principles that govern the coming-into-being of a new biological species in nature. What follows is an attempt to comprehend that living process, to listen to the resonance that still hums beneath the streets and squares of old

Constantinople, and to recognize in the equations of morphic resonance not cold abstractions but the very essence of a living organism.

Stand on the shore of the Bosporus in the year 395. The city is still called New Rome. Senators speak Latin in the marble halls of the Senate house; the legions march under the old standards; the emperor is addressed as Augustus and Imperator. Yet already something is stirring. In the chanceries as scribes write more and more in Greek instead of Latin. In the city's churches, the liturgy is chanted in the tongue of Plato and the Gospels; in the workshops of architects, sketches are being made for churches whose central domes will one day float above the city as if suspended from the heavens. Five centuries later, in the year 900, the same city is called Constantinople, the City of Constantine, and although its people still call themselves Romans, they speak Greek as naturally as they breathe. They worship beneath mosaics of Christ Pantokrator, whose eyes follow the soul across the nave, and their emperor is crowned not with the laurel of the Caesars but with the stemma and loros of a cosmic sovereign who stands between God and man. The Corpus Juris Civilis of Justinian has become the Basilika of Leo the Wise and the ancient Roman basilica plan for churches has given way to the Byzantine cross-in-square. The old Roman Stoic restraint has flowered into an icon-rich mysticism that floods the senses with the gleam of gold and the rich odor of incense. How did this happen? Not by invasion alone, though invaders came; not by imperial decree alone, though emperors decreed much; not by the slow diffusion of ideas, for ideas do not diffuse with such precision and speed. Something else was at work: a morphic field of influence that inherited the full weight of Rome and then, slowly, lovingly, transfigured it into something new.

Every new beginning comes with a small difference. In biology, a new species arises when a population is separated from its parent stock and a slight variation appears—perhaps a different beak shape or a different mating call. The initial variation is tiny, almost invisible. Yet if the separating barrier endures, the daughter population begins to resonate more strongly with its own past than with the past of its parent, and in time the two fields diverge and a new species comes into being. So it was with Byzantium. The separation began quietly in 395, when Emperor Theodosius the Great divided the empire between his two young sons. The West received Honorius; the East received Arcadius. The division was meant to be temporary, a political convenience. Yet geography, wealth, and language conspired to make it lasting. The East was richer, more defensible, and more Greek. The West was poorer, more exposed to barbarian invasion, and more Latin.

A small perturbation—designated by ε in the mathematics of morphic resonance—was introduced into the Eastern field. At first the difference was almost nothing. The court in Constantinople still used Latin for official documents; the army still swore the old Roman oaths; and the bishops still looked to Rome as the seat of Peter. But beneath the surface the new field was gathering itself. Greek was the language of scripture and philosophy, the language people actually spoke from Antioch to Thessaloniki. The churches of the East had always loved splendour and symbolism. The emperor had long been seen, in the Greek-speaking provinces, as the representative of God on earth. These were not innovations. They were latent possibilities within the Roman field itself, possibilities that now began to resonate more strongly in the East than in the West.

For a long time the parent field remained dominant. Justinian, reigning in the sixth century, felt its pull with extraordinary intensity. He sent his armies to reconquer Italy, North Africa, and southern Spain in a heroic effort to make the Mediterranean world whole again; he commanded his lawyers to compile the Corpus Juris Civilis in Latin, the language of old Rome; he built Hagia Sophia, whose magnificent dome was the largest in the world for a thousand years. Yet even in these acts of Roman restoration we can see the daughter field beginning to assert itself. The Code, for all its Roman precision, simplified and Christianized the law. The dome of Hagia Sophia, for all its Roman engineering, was a Greek and Eastern dream of heaven brought to earth.

Then came the great geopolitical catastrophes. They plague devastated the East in the sixth century. The Arab Muslim conquests overwhelmed Egypt, Syria, and North Africa. The empire lost much of its territory. In the West, the last emperor was deposed in 476; the Roman parent field fragmented into a hundred barbarian kingdoms. In the East, the daughter field was thrown back upon itself. Greek became the language of survival. The old provinces were replaced by military themes. In the 7th century the heroic Emperor Heraclius adopted the title of "Basileus, " the old Greek word for king or emperor. The Latin echo grew fainter, not because it was hated, but because it was no longer needed. This fading is described in the Annex by an exponential decay term in Equation 2. This does not represent active rejection, but rather a gentle forgetting. The daughter culture of Byzantium does not turn her back on her Roman mother. She simply ceases to hear her voice clearly.

As the parent's voice receded, the daughter's own voice grew stronger. Practices that worked—Greek administration, domed churches, the theme system, the veneration of icons—were repeated, refined, and reinforced. Each repetition strengthened the field. The Ecloga of Leo III in the eighth century was simpler, more Christian, more Byzantine, and easier to apply than Justinian's Code. The cross-in-square church, intimate and radiant with icons, became the natural way to build. The emperor became not merely first citizen but God's vicegerent, crowned in a ceremony heavy with cosmic symbolism. Crises accelerated the process. The Arab invasions stripped away the old Middle Eastern and North African provinces and concentrated the empire's identity in new key regions: Anatolia, the Aegean islands, the Balkans, and the Black Sea.

The Iconoclastic controversy of the eighth and ninth centuries was a terrible trial of great importance. For a century the empire tore itself apart over profound questions of theology having to do with the nature of divine imagery and worship. When the storm passed and the use of icons in religious rituals was restored in 843, the field incorporated them as essential to its own nature. Orthodoxy triumphed, not as a return to the past but as a discovery for the future. By the ninth century, under the Macedonian dynasty, the daughter field had crossed the threshold of autonomy. The overlap with classical Rome—measured in the mathematics of morphic resonance by the integral of shared forms across the three-dimensional temporal manifold—had fallen beneath the critical value θ_c. The Basilika of Basil I and Leo VI was no longer a translation of Justinian but a new legal framework resonating primarily with centuries of Byzantine practice. The Macedonian Renaissance was not a revival of classical antiquity; it was the first full flowering of a field that now resonated with itself. Its ambitious political, military, and cultural expansion was sharply different from that of Justinian, with the revived empire sending Orthodox missionaries to convert the Russians in the north, and armies to subdue Bulgaria in the west and Armenia in the east. The empire became by far the most powerful state in Europe and the Middle East by the early 11th century.

For centuries the Byzantine world lived in a state of superposition. A law could be both Roman and Byzantine at once; a church could have the proportions of a basilica and the soul of a dome; an emperor could be hailed as Augustus and yet kneel before Christ Pantokrator. This creative entanglement is captured in the bifurcation equation of morphic resonance. Every cultural trait existed as a weighted combination of inherited Roman form, emerging Byzantine novelty, and a subtle interference term that produced hybrids neither parent nor daughter could have foreseen. The dome on pendentives is the perfect embodiment of this superposition. It is Roman in its concrete and engineering, Greek in its harmonious mathematical design, Eastern in the symbolism of the vault of heaven, and Christian in its liturgical function. Justinian's architects did not invent it ex nihilo; they tuned into a possibility that had been gathering strength in the field for centuries. When Hagia Sophia rose in 537, the builders themselves are said to have fallen to their knees, believing that angels had guided their hands. They were not wrong. They had been guided—by the resonance of countless previous domes, real and potential, stretching back through Armenian churches and Persian iwans to the very idea of heaven itself.

Speciation is never complete until the fields separate cleanly. For centuries East and West continued to share the same Christ, the same creeds, and the same saints. Yet their morphic fields had clearly diverged. In the West the field hardened around Catholic religious law, the hierarchy of the Vatican, and the primacy of the Popes. In the East it opened into mystery, hesychasm, and the sense that the liturgy itself was an icon of the Kingdom of Heaven. When the legates of Pope Leo IX laid the bull of excommunication on the altar of Hagia Sophia in 1054, they were not causing the schism. They were acknowledging a splitting of morphic fields that had already been accomplished. The sack of Constantinople by the Fourth Crusade in 1204 was the final, brutal confirmation of this division. Latin knights desecrated the holiest altar in Christendom, installed a foreign emperor on the throne of St. Constantine, and looted priceless relics such as the True Cross and the burial shroud of Christ. The repulsive gradient in the partition equation had done its work. Two cultural manifolds now existed, topologically distinct. Resonance between them became mere background noise.

Even half a millennium after the conquest of Constantinople by the Ottoman Turks, the Byzantine morphic field is not dead. Walk into any Orthodox church today—whether in Athens, Moscow, or a quiet village in the Carpathians—and you will feel it. The dome above your head, the icons gazing down from the vaults and walls, the chant that seems to come from nowhere and everywhere at once—these are not relics. They are the habits of a field that has been singing the same sacred melodies, with slight variations, for fifteen centuries. Even in the secular West we can hear its echo. Our legal systems still carry the resonance of Justinian and his Code. Our sense of sacred kingship, however attenuated, descends from the Byzantine Basileus. Our very idea that a civilization can endure by continual metamorphosis rather than rigid preservation—that idea was born in Constantinople between the seventh and ninth centuries, when a wounded empire discovered that it could survive and thrive by being born again.

Since civilizations are indeed governed by morphic fields, then history is not a meaningless succession of events. It is a process of morphogenesis on a very large scale. Forms do not arise by chance and blind selection alone. They are drawn out of possibility by resonance with what has gone before. Memory is not confined to brains or libraries; it resides in extensive fields that embrace entire cultures and reach across centuries. Evolution, at this level, is Lamarckian: what a civilization practices with devotion, it preserves and passes on directly through resonance. The habits of precision and balance in Roman law were practised for centuries; they entered the field and were inherited by Byzantium, transfigured but not lost. The habit of seeing the emperor as God's servant and representative was practiced with ever greater intensity; it became the organizing centre of a new political theology. A civilization's collective consciousness is nothing more than the self-resonance of its cultural field when it grows strong enough to feel its own past as a living presence. When an eleventh-century monk stood beneath the dome of Hagia Sophia and felt the boundaries between earth and heaven dissolve, he was not indulging in a mystical revery. He was participating in the Byzantine morphic field at full strength, a field in which Roman engineering, Greek philosophy, and Christian sacraments had all been gathered into a single, resonant chord.

We have been taught to mourn the fall of Rome. We should instead celebrate the birth of Byzantium. From a minute perturbation in the fourth century—a slight drift toward Greek speech, a slight preference for domed space, a slight intensification of imperial theology—a new morphic field detached itself from its parent, faded from the parent's influence, discovered its own attractors, and stabilised into one of the longest-lived and most radiant civilizations humanity has ever known. That is how all living forms come into being, whether biological or historical. The past does not push from behind. It pulls from ahead. The future is not built brick by brick; it is remembered into existence. If we listen carefully, across the thousand years that separate us from the golden mosaics of the Macedonian Renaissance, we can still hear the daughter field singing—quietly, persistently, in every dome that rises above an Orthodox church, in every legal code that seeks balance and continuity, in every intuition that a society can endure by changing beautifully and sustainably. The Byzantine field is not dead. It lives and still speaks to anyone who has ears to hear.

Mathematical Annex

(Glossary of Symbols and Concepts Used in the Essay)

Core Field Quantities

$M_P^i(t)$

The morphic field of the parent civilization (late antique Rome) at time t. The index i labels different cultural traits (language, law, architecture, imperial ideology, etc.).

$M_D^i(t)$

The morphic field of the daughter civilization (Byzantium) at time t.

ϕ_P^i, ϕ_D^i

Specific realised forms or patterns (e.g., a basilica, a Latin legal text, the title "Augustus") within the parent or daughter field.

t_{split}

The moment of effective speciation (historically somewhere between 395 CE and c. 537 CE).

ε

A very small number representing the magnitude of the initial mutational perturbation that seeds the daughter field (the tiny Greek-speaking, dome-leaning, theocratic drift in the East).

d_{PD}

"Form distance" in cultural trait space between parent and daughter fields at any moment.

Equations and Their Interpretation

(1) Daughter Field Initialization

$$M_D^i(t_{\text{split}}) = M_P^i(t_{\text{split}}) + \varepsilon\, \Delta M_{\text{mutation}}^i(t_{\text{split}}^k)$$

$\varepsilon\, \Delta M_{\text{mutation}}^i$ – The tiny initial difference (e.g., greater use of Greek in eastern documents, early experiments with centralised church plans) that distinguishes Byzantium from Rome at the moment of separation.

(2) Divergence Kernel (fading parental influence)

$$K_{D\leftarrow P\,j}^i = G_{\text{div}}(d_{PD}^2)\, \delta_j^i\, e^{-\lambda(t - t_{\text{split}})}$$

$K_{D\leftarrow P\,j}^i$ – The time-decaying influence that the Roman past continues to exert on the emerging Byzantine present.

$G_{\text{div}}(d_{PD}^2)$: a Gaussian or similar function of form distance

λ: decay rate of Roman influence

$e^{-\lambda(t-t_{\text{split}})}$: exponential fading of the parental echo

(3) Daughter Field Evolution

$$\frac{\partial M_D^i}{\partial t} = \kappa_D\, K_{D\,j}^i \phi_D^j + \kappa_{D\leftarrow P}\, K_{D\leftarrow P\,j}^i \phi_P^j + \eta\, \xi^i(t^k)$$

Rate of change of the Byzantine cultural field.

$\kappa_D K_{D\,j}^i \phi_D^j$: self-resonance term (growing Byzantine habits)

$\kappa_{D\leftarrow P} K_{D\leftarrow P\,j}^i \phi_P^j$: fading Roman pull

$\eta\, \xi^i(t^k)$: stochastic innovation

(4) Speciation Threshold

$$\mathcal{O}_{DP}(t) = \frac{\int M_D^i M_P^i \, d^3\tau}{\sqrt{\int M_D^j M_D^j \, d^3\tau \cdot \int M_P^k M_P^k \, d^3\tau}} < \theta_c$$

$\mathcal{O}_{DP}(t)$: normalised overlap integral over the three-dimensional temporal manifold
θ_c: critical threshold (reached ≈ 870–900 CE); below this, Byzantium is morphologically autonomous

(5) Form Bifurcation

$$\phi_D^i = \sqrt{1-p}\,\phi_P^i + \sqrt{p}\,\phi_{\text{novel}}^i + \sqrt{p(1-p)}\,\Phi_{\text{entangle}}^{ij}\phi_{P\,j}$$

$p \in [0, 1]$: bifurcation parameter (Roman → Byzantine shift)
The three terms represent inherited Roman form, emerging Byzantine novelty, and their creative entanglement

(6) Morphic Cloud Partition

$$\frac{d^2\gamma_D^k}{d\lambda^2} = \kappa_D\,K_{Dl}^k\,\delta\phi_D^l - \kappa_{D\leftarrow P}\,\nabla^k\mathcal{O}_{DP}$$

Final topological separation: attractive self-resonance within Byzantium combined with a repulsive gradient away from the collapsing Roman attractor (visible symptoms: 1054 Schism and 1204 Sack).

XVI

The Morphic Cloud and Cultural Revolution: the Renaissance in Florence

The strange and beautiful idea at the core of this book is that past forms, once they have come into being, do not altogether vanish. They leave behind them something like a trace, a memory inherent in nature itself, which makes it easier for similar forms to appear again in the future. This trace is not stored in some celestial archive, nor is it carried only in books and buildings, though books and buildings can be its instruments. It is held, rather, in a field—a field embedded in the three-dimensional structure of time. It is a field to which members of a society, or a species, are attuned by a process akin to resonance. These fields are called morphic fields, and the process by which they exert their influence morphic resonance.

When we turn to the great awakenings of history, those periods when a civilisation seems suddenly to remember what it once knew and to leap forward into new possibilities, the idea of morphic resonance offers a way of seeing that is neither purely materialistic nor merely mystical, but something in between—an extended science of connection and memory. Few awakenings have been more vivid than the Renaissance in Florence, that small city on the river Arno which, between the early years of the fifteenth century and the first decades of the seventeenth, gave rise to an extraordinary flowering of painting, sculpture, architecture, philosophy, science, and poetry. How did this happen, and happen so quickly, in a city of scarcely a hundred thousand inhabitants?

Let us approach the question not through the usual lists of economic causes and political patrons—though these mattered—but through the living atmosphere of the place itself, the field of collective habit and possibility that surrounded every Florentine from cradle to grave. Imagine this field as a kind of cloud, invisible yet palpable, embedded in three-dimensional time, hovering over the narrow streets, the workshops, the palazzi, the churches, and even the river. Into this cloud were woven the habits of seeing, thinking, making, and believing that had taken shape over generations. And then, quite suddenly, the cloud began to change its shape, to brighten, to expand, to resonate with patterns that had lain dormant for a thousand years or more.

In the beginning, the morphic cloud of medieval Florence was dense with the forms of Gothic Christianity. It contained the habits of mind that built the tall lancet windows of Santa Croce, the habits of hand that painted gold-backed Madonnas, the habits of thought that placed God at the centre of an ordered cosmos and man as a humble pilgrim within it. These habits were strong and self-reinforcing, tuned by centuries of repetition. A young painter entering the workshop of a master like Giotto or Taddeo Gaddi would find himself surrounded by this field, and the field would make it natural—almost inevitable—that he paint in the established manner. The past was present, guiding the brush.

Yet even within this dense cloud there were faint tremors of something else. Merchants and scholars returning from Constantinople brought back manuscripts of Plato and Plotinus. Scholars in the monasteries copied Vitruvius and dreamed of Roman domes. Byzantine painters, refugees from the falling Empire, showed that figures could be modelled with light and shadow rather than outlined in black. These were minority patterns, weak resonances, yet they were there, trembling on the edge of visibility.

Then, in the early decades of the fifteenth century, the cloud began to shift. The shift was not gradual; it was more like the moment when a supersaturated solution suddenly crystallises, or when a tuned string begins to sing in sympathy with another. A new pattern appeared with astonishing clarity: the pattern of linear perspective. Filippo Brunelleschi, standing inside the unfinished Duomo, demonstrated to his friends how parallel lines could be made to converge toward a vanishing point, how the pavement of the Baptistery could be rendered as a geometrical grid receding into depth. Almost overnight, the possibility of representing three-dimensional space on a two-dimensional surface leapt into the common field.

How are we to understand the speed of this leap? In the usual histories we read of Brunelleschi's experiments, of his friendship with Donatello, of the competition for the bronze doors of the Baptistery. These are the surface events. Beneath them something subtler was taking place. The moment Brunelleschi grasped the principle of perspective, the morphic field of Florentine art was altered. The new pattern was now available, not only to those who had watched Brunelleschi's demonstration, but to everyone attuned to the same field. Masaccio, a young painter barely out of his teens, walked into the Brancacci Chapel and began to paint the Tribute Money with a spatial coherence that would have been unthinkable twenty years earlier. Piero della Francesca in distant Arezzo, Paolo Uccello in his obsessive nights of drawing, Leon Battista Alberti writing his treatise On Painting—all were drawn into the same resonance, as if the new habit of seeing had been waiting, fully formed, for the right moment to declare itself.

This is the signature of morphic resonance: the sudden appearance of a new form, followed by its rapid spread, followed in turn by its stabilisation as the new normal. We see the same rhythm again and again in Renaissance Florence. Donatello carves the marble David with a sensuous freedom no medieval sculptor had permitted himself; within a generation the contrapposto pose and the celebration of the nude human body have become commonplace. Ghiberti finishes the Gates of Paradise; suddenly bronze relief can tell a story with the fluency of painting. Botticelli dreams of Venus rising from the sea, and mythological themes, banished for centuries, flood back into Christian Florence as though they had never been away.

And it was not only in the visual arts. The same cloud that shaped the eyes of painters shaped the minds of scholars. When Marsilio Ficino sat down to translate Plato into Latin under the patronage of Cosimo de' Medici, he was not working in isolation. The Platonic pattern—the sense that the visible world is a reflection of eternal ideas, that the soul is akin to the divine, that beauty is a path toward the Good—had been growing stronger in the field for decades. Petrarch had felt it, Boccaccio had felt it, even Dante had felt it. Ficino's translations did not create the pattern; they released it. Once released, it resonated through lecture halls, palaces, and convents. Michelangelo would later say that the sculpture already existed inside the block of marble and that his task was merely to remove the superfluous stone; in the same spirit one might say that the Platonic Renaissance already existed inside the Florentine cloud, waiting for the right hands and minds to set it free.

The Medici, of course, played their part. They were not the cause of the new field but its facilitators. By gathering painters, sculptors, poets, and philosophers under their roof, by paying for libraries and academies, by commissioning works that pushed artists beyond the conventional, they increased the intensity of the resonance. A field is strengthened by the number and activity of the systems attuned to it. When Leonardo, Michelangelo, and

Raphael were all alive in Florence at the same time—though they were often quarrelling and competing—the morphic field of art must have glowed like iron in a forge.

Nor should we neglect the role of the city itself. Florence was small enough for ideas to travel quickly on foot. A discovery made in Brunelleschi's workshop one morning could be discussed in the Platonic Academy that evening and tried out in a painter's studio the next week. The cloud was dense, intimate, highly responsive. And because Florence was a republic—or liked to think of itself as one—there was a premium on individual excellence, on standing out, on doing something no one had done before. Competition and collaboration fed each other, and both fed the morphic field.

Yet the field was never sealed within the city walls. Trade routes, embassies, wandering scholars, and printed books carried Florentine resonances outward. When Albrecht Dürer in Nuremberg began to draw with a new precision of line and volume, when Raphael in Rome painted the School of Athens with its perfect perspective, when even Shakespeare in distant London caught echoes of Italian humanism, they were not merely imitating what they had seen or read. They were tuning in to a field that had grown strong enough to be felt across mountains and seas.

There is a further mystery here, one that touches on the nature of time itself. Why did the Renaissance happen when it did? Why in the fifteenth century and not the twelfth, or the eighteenth? Part of the answer lies in external events—the fall of Constantinople driving Greek scholars westward, the invention of printing making texts multiply, the wealth generated by banking and the wool trade. Yet these are triggering conditions rather than ultimate causes. Beneath them lies a deeper rhythm. Patterns that had been dormant since the collapse of the classical world were growing stronger again, not because copies of Plato were suddenly available—copies had existed in monasteries for over a thousand years—but because the morphic field of Western culture was ready to resonate with them once more.

Imagine the great forms of the past—Greek philosophy, Roman engineering, the human-centred art of antiquity—not as dead and gone but as sleeping possibilities within the extended mind of Europe. For a thousand years the Western Christian field had kept them suppressed, or rather had overlaid them with a different pattern. Then, slowly at first and then with gathering speed, the older resonances reasserted themselves. Florence, by historical accident, became the place where the new tuning fork was struck.

The note sounded there first, but it was a note the whole continent was waiting to hear.

What became of the Florentine cloud when the great period ended? It did not vanish. It dispersed, diffused, and was transformed. The intensity that had been concentrated in one small city spread outward, carried by artists who travelled, by books that were printed, by pupils who studied and then taught elsewhere. The field of European culture as a whole grew richer, more complex, more various. Mannerism, Baroque, Neoclassicism—all would draw on the resonances first stabilised in Florence. Even today, when we stand before the David or walk beneath Brunelleschi's dome, we are still in the grip of that field. It has not lost its power to move us, because the habits of seeing and feeling it established have become part of our inheritance and nature.

The idea of morphic resonance does not explain everything. It does not tell us why Cosimo de' Medici loved Plato, or why Michelangelo suffered as he did, or why the Black Death left Florence wealthy rather than ruined. But it gives us a way of seeing these events as something more than a chain of material causes. It suggests that human creativity is not an isolated phenomenon, the product of separate brains in separate skulls, but a process of attunement to fields that stretch beyond the individual and the present moment. In the Florence of the Renaissance, we see this process at its most luminous: a society suddenly remembering what it had forgotten, and in remembering, becoming something new.

Perhaps that is the deepest gift of those years on the Arno. They remind us that cultures, like organisms, have memories; that the past is not dead but lives on in fields of influence we are only beginning to understand; and that from time to time, when conditions are right, a society can wake from a long sleep and rediscover, with astonishment and joy, capacities it did not know it possessed. The cloud over Florence in those centuries was bright with possibility, and some of its light still reaches us today.

XVII

Morphic Resonance and the American Revolution

The years between 1775 and 1783 witnessed one of the most startling transformations in human history. A scattered collection of provinces that had thought of themselves as British for a century and a half suddenly, almost overnight in historical terms, declared themselves a new nation, fought a war to make the declaration good, and then proceeded to invent an entirely new form of government that had no precedent at that scale. Farmers in Massachusetts and planters in Virginia, merchants in Philadelphia and back-country riflemen in the Carolinas, all discovered within months that they shared a common political nature that overrode older loyalties of province, religion, and class. Pamphlets printed in one city were read aloud in taverns a thousand miles away and produced the same emotional and intellectual effect, as though the colonists were not separated by weeks of slow travel but linked by some invisible medium that carried ideas faster than horses or ships.

Conventional history explains this through a shared sense of grievance toward Britain, brilliant political and military leadership on the side of the Americans, and the rapid spread of ideas. Yet the speed, the simultaneity, and above all the coherence of the change remain mysterious. How did thirteen legislatures, often quarrelling among themselves, arrive almost simultaneously at the conviction that independence was not only desirable but inevitable? Why did the same phrases—"the rights of Englishmen," then "the laws of nature and of nature's God"—appear simultaneously in documents from New Hampshire to Georgia? Why did militia companies spring up with the same drill and the same determination in places that had no direct communication?

Rupert Sheldrake's hypothesis of morphic resonance offers a way of thinking about such events that is neither mystical nor merely metaphorical. Nature, he proposes, is habitual. Systems of every kind—crystals, molecules, organisms, societies—are influenced by traces left by previous systems of the same kind. These traces are not stored in matter or energy as we usually understand them but in fields that extend across space and time, and that exert their influence through resonance: the more similar a present system is to past systems, the stronger the influence. A new tune is easier to learn if many people have already played it; a new chemical compound crystallises more readily if the same compound has crystallised elsewhere; a new social form stabilises more quickly if similar forms have existed before.

In biology Sheldrake first applied the idea to the appearance of new species or new behaviors. Here we extend it to the sudden appearance of a new political system through the political equivalent of speciation. Seen from this perspective, the American Revolution was a speciation event in the body politic. From one relatively homogeneous colonial organism there separated, irreversibly, two distinct forms: the continuing British imperial polity and a new American republic organized around written constitutions, popular sovereignty, and the deliberate rejection of hereditary rule. The mathematics that describes this process is based upon the equations used in this book to analyze biological speciation (see Equations 1–6). Suitably adapted to political transformation, is they are presented in full in the Mathematical Annex at the end of this chapter. What follows is the living story of how a morphic field of revolution came into being, grew, crossed a critical threshold, and finally stabilized a new pattern of collective life that still thrives and resonates today.

The background field of colonial British America was stable for generations. It was a field of deference: deference to the Crown, to Parliament, to governors and councils appointed from London, to a social hierarchy headed by a distant king and a home-country aristocracy. This field was not merely ideological; it was habitual, embodied in countless daily practices—raising a toast to the King's health, dating documents by the regnal year, accepting the presence of red-coated soldiers as natural. The field was reinforced every time a colonist paid a customs duty, attended an Anglican service, or read a London newspaper. It was a morphic field in the strict sense: the more colonists performed these acts, the more natural the acts felt, and the more probable they became for others.

Yet fields can be stretched. The Stamp Act of 1765, the Townshend duties, the Quartering Act, the closure of Boston harbour—each new assertion of parliamentary power pulled at the habitual fabric. Most of the time such tensions are repaired; grievances are negotiated, concessions granted, and the old field reasserts itself. Yet sometimes the stretch becomes critical. A small innovation appears that cannot be reabsorbed: the claim that representation is not a privilege granted by Parliament but a natural right inseparable from taxation. Once that idea is uttered aloud—and it was uttered many times in the decade before 1775—it begins to resonate. Every new utterance makes the next one easier. The field begins to tremble.

The decisive moment of splitting is difficult to fix with precision, but let us take the early morning of 19 April 1775, when British regulars marched out from Boston to seize colonial arms and the farmers of Massachusetts turned out to oppose them. Something irreversible happened on the road back from Concord. The muskets had been fired; blood had been shed; the habitual field of deference had cracked. At that instant an American daughter-field began to separate from the British parent (see Equation 1). It was almost identical to the old field at first—the same language, the same churches, the same towns—but it carried a tiny mutational difference: the possibility that armed resistance to the King's troops could be legitimate.

From that moment on the two fields began to diverge. The parent field—British imperial identity—continued to exert influence, but its pull weakened with every new act that increased the distance between colonial practice and British expectations (see Equation 2). News of Lexington and Concord reached Charleston, South Carolina, weeks later, yet the response was immediate: militia companies formed, committees of safety organised, and royal officials fled. The influence was not travelling by ship or wagon; it was resonating across a morphic field that already contained the latent form of resistance.

As the daughter-field grew, it began to develop its own internal laws. Every new committee of correspondence, every new boycott association, every new militia drill strengthened the field and made the next such act more probable. At the same time, news of British retaliation—the burning of Falmouth, the hiring of Hessian mercenaries—acted as a repulsive force, pushing the colonial consciousness farther from the old habits (see Equation 3). And everywhere there was creativity: printers turning out broadsides, preachers finding republican themes in the Book of Samuel, and ordinary men and women discovering that they could govern themselves in town meetings without a lord or royal official.

By January 1776 the daughter-field was strong enough to support a phenomenon that still astonishes historians: the publication and instantaneous continent-wide adoption of Thomas Paine's Common Sense. In a few weeks 120,000 copies were sold in a population of two and a half million—an information explosion comparable in its day to "breaking the internet" today. People who had never thought systematically about independence found the idea suddenly natural and self-evident. The morphic field had reached a density where a single clear statement could flip thousands of individual habits almost simultaneously.

Seven months later, in July 1776, the Continental Congress declared independence. This was not the beginning of the process; it was the public acknowledgement that the daughter-field had already grown stronger than the parent. The Declaration itself became a new resonant object: every public reading, every copy posted on tavern walls, strengthened the field further. And the field was not confined to the present; it reached backward to the English Civil War, to Cromwell's commonwealth, to the ancient Israelite commonwealth in the Book of Judges, and forward to an imagined future in which free republics would cover the earth. Past and future both contributed to the stability of the new form.

Yet declaration is not the same as irreversible separation. Loyalism remained strong in many areas; the war dragged on for five more years. There were moments—Valley Forge, Camden, the mutinies of 1781—when the new field seemed about to collapse back into the old. What prevented collapse was the gradual fall in the overlap between the two fields (see Equation 4). As more people swore oaths to their home states rather than to the distant King, as more soldiers wore continental uniforms instead of provincial colors, as more printers dated their newspapers "In the Year of Our Lord and the First Year of American Independence," the normalized overlap between revolutionary habits and imperial habits dropped lower and lower. When it crossed a critical threshold—somewhere between the French alliance of 1778 and the victory at Yorktown in 1781—the speciation became irreversible (see Equation 4). A new political species had been born in the form of the American Republic.

The new American form was not a pure novelty. It carried within it large portions of the British parent field: trial by jury, representative assemblies, the common law, the English language itself. But these elements were recombined with radical innovations—no hereditary monarchy, no established church, no titles of nobility, and sovereignty residing in "We the People" (see Equation 5). Crucially, the new form contained an entanglement term that bound thirteen jealous republics into a single field strong enough to survive the centrifugal forces of localism (see Equation 5). The Constitution of 1787 was the final stabilization of this entangled superposition.

After 1783 the two fields—British and American—parted company almost completely (see Equation 6). A British subject in London and an American citizen in Philadelphia now inhabited different regions of the temporal manifold. Their respective morphic clouds no longer overlapped significantly. Yet faint resonances persisted: the American Constitution borrowed heavily from British practice, and British reformers in the nineteenth century would look across the Atlantic for inspiration. The entanglement had not vanished; it had merely stretched across an ocean and a revolution.

Seen in this light the American Revolution is not an isolated event but an archetype of political speciation. The French Revolution resonated with the American field and amplified it, then quickly diverged into its own terrible and glorious form. The Latin American revolutions of the early nineteenth century drew strength from both. The Haitian Revolution, the Greek Revolution, the revolutions of 1848—all were daughter-fields seeded from earlier ones, each with its own balance of inheritance, novelty, and entanglement.

The implications reach far beyond the eighteenth century. If morphic resonance operates in history, then revolutions are never merely the product of local material causes, however important they may be. They are also influenced by fields that belong to the past of their own kind and, in some sense, to the future attractors of human possibility. The American revolutionaries felt this intuitively. They spoke constantly of posterity, of setting an example that would be imitated by oppressed people everywhere, of being watched by the eyes of future ages. Jefferson wrote that the Declaration was meant to be "an expression of the American mind" and also "to place before mankind the common sense of the subject, in terms so plain and firm as to command their assent, and to justify ourselves in the independent stand we are compelled to take." In the language of morphic resonance he was describing the deliberate strengthening of a field that would make future acts of self-determination easier wherever similar conditions arose.

The mathematics presented in the Annex is not ornamental. It is an invitation to consider the American Revolution and other political revolutions as phenomena that can be described by the mathematics of morphic resonance developed in this book, specifically by the adapted equations for speciation (see Equations 1–6). Historians could assemble dated sequences of revolutionary acts—pamphlet publications, militia enrollments, legislative votes, newspaper editorials—and feed them into the equations. Where the model predicts rapid decline in the overlap between old and new fields and history confirms it, we have evidence that something beyond local causation was at work. Until such tests are performed we may simply stand in wonder at what happened between the shot heard round the world and the silence that followed Cornwallis's surrender at Yorktown.

In those few years a new political species came into being—suddenly, coherently, against very long odds. A morphic field, seeded by grievance, nourished by courage, entangled by shared sacrifice, and stabilized by resonance across time, brought forth a nation conceived in liberty and dedicated to the proposition that human beings are capable of governing themselves. The field still hums across the centuries, making certain habits of self-government feel natural to those who have never known any other way, and offering a distant but real encouragement to those who are only now discovering that such habits are possible.

Mathematical Annex: Equations of Political Revolution

(Adapted from the morphic-speciation formalism to the American Revolution 1775–1787; all fields and integrals are defined on Dr. Gunther Kletetschka's 3+3-dimensional space–time manifold)

1. Daughter-Field Initialisation

$$M^i_R(t_{\mathrm{split}}) = M^i_B(t_{\mathrm{split}}) + \varepsilon\,\Delta M^i_{\mathrm{mutation}}(t_{\mathrm{split}})$$

Seeds the revolutionary American morphic field from the colonial British parent field at the instant of political splitting. The new field is initially almost identical to the old, differing only by a small innovative perturbation (e.g., the legitimacy of armed resistance).

2. Divergence Kernel

$$K^i_{R\leftarrow Bj} = G_{\mathrm{div}}(ds^2)\,\delta^i_j\,e^{-\lambda(t-t_{\mathrm{split}})}$$

Governs the rapid fading of imperial British influence on the emerging republican form. The Gaussian factor (using the squared temporal-ideological distance ds^2) and exponential decay ensure that parental pull collapses as ideological and behavioural distance grows.

3. Daughter-Field Evolution

$$\frac{\partial M^i_R}{\partial t} = \kappa_D K^i_{Rj}\phi^j_D + \kappa_{R\leftarrow B} K^i_{R\leftarrow Bj}\phi^j_B + \eta\,\xi^i(t)$$

Master equation describing the growth of revolutionary habits through (i) self-reinforcement within the daughter field, (ii) residual (usually repulsive) influence from the British field, and (iii) stochastic bursts of political creativity.

4. Speciation Threshold

$$O_{RB}(t) = \frac{\int M_R^i M_B^i \, d^3t}{\sqrt{\int M_R^j M_R^j \, d^3t \cdot \int M_B^k M_B^k \, d^3t}} < \theta_c$$

Normalised overlap integral between revolutionary and imperial morphic fields. When O_{RB} falls below the critical value θ_c ≈ 0.37, political speciation becomes irreversible; the new republic is autonomous.

5. Form Bifurcation

$$\phi_D^i = \sqrt{1-p}\,\phi_P^i + \sqrt{p}\,\phi_{\text{novel}}^i + \sqrt{p(1-p)}\,\Phi_{\text{entangle}}^{ij}\phi_P^j$$

Expresses the daughter form as a weighted superposition of parental inheritance, novelty, and contextual entanglement.

6. Morphic Cloud Partition

$$\frac{D^2\gamma_D^k}{d\lambda^2} = \kappa_D K_{Dl}^k \,\delta\phi_D^l - \kappa_{D\leftarrow P}\nabla^k O_{DP}$$

Separates daughter developmental paths via attractive self-resonance and repulsive parental overlap gradients.

XVIII

Abraham Lincoln as the Embodiment of Jungian Archetypes

Certain rare figures in history seem to step out of the collective morphic field of humanity, to embody the heroic past as they labor in the present moment to create the future. Abraham Lincoln is one such figure. He is not merely a president who happened to live through a great crisis; he is a nodal point in the extended mind of the American people, and perhaps of Western civilisation itself. Through the lens of Carl Jung's archetypal psychology and the theory of morphic resonance – the idea that memory is inherent in nature, that similar patterns of form and behaviour are influenced by fields that connect past and present across time and space – Lincoln appears less as an isolated individual and more as a living constellation of eternal images. In this sense, he is the Hero, the Wise Old Man, the Caregiver, the Everyman, and the Ruler. These images did not originate with him, yet in him they found an extraordinary focus of expression, and from that focus they continue to radiate, organizing the political and imaginative life of a nation down to our own day.

Let us begin our account when the field itself seems to have begun to gather intensity: in the 1850s, that decade when ominous clouds thickened in the American political sky. The Whig Party, which had carried the banner of national economic development since the days of Henry Clay, was coming apart along a fault-line that was at once moral and material. Slavery, long contained by uneasy compromises, was expanding territorially again. The Kansas-Nebraska Act of 1854 tore open the Missouri Compromise and replaced it with the doctrine of "popular sovereignty," a phrase that sounded democratic but in practice handed the decision about human bondage to armed settlers who rode into Kansas with Bibles in one hand and Sharps rifles in the other. The result was Bleeding Kansas, a miniature civil war that killed two hundred people and shattered the last illusions of sectional harmony within the old parties.

In biological terms, what happened next resembles sympatric speciation: a new species arising within the same geographical space as its parent because a portion of the population can no longer reproduce – in this case, politically – with the rest. The anti-slavery Conscience Whigs, the Free-Soilers, and a scattering of northern Democrats found themselves unable to ally with the Cotton Whigs and the doughface Democrats who were willing to appease the slave states. On 20 February 1854, in a small schoolhouse in Ripon, Wisconsin, a former Whig named Alvan Earle Bovay proposed that they call the new movement "Republican," a name deliberately chosen to evoke Jefferson's old party of the early 1800s and its commitment to equality under the laws of nature. Five months later, on 6 July 1854, ten thousand people gathered under the oaks of Jackson, Michigan, and formally brought the Republican Party into existence. A morphic field had crystallised.

Onto that newly forming field stepped Abraham Lincoln. He was forty-five years old, a lanky, self-taught lawyer from Springfield, Illinois, who had served one undistinguished term in Congress and then returned to the prairie to practice law among farmers and small-town merchants. He was not the most famous anti-slavery voice – Seward, Sumner, and Chase all outshone him in reputation – yet something in the field was already beginning to organize itself around him. When he spoke at Peoria in October 1854 against the Kansas-Nebraska Act, and again at Springfield in 1858 with the words "A house divided against itself cannot stand," people felt that an older pattern was reasserting itself: the pattern of the Hero who confronts the dragon of chaos.

Jung described the Hero as the archetype of transformation. The Hero is born in lowly circumstances, undergoes trials, descends into the darkness, slays or redeems a beast or monster, and returns with a boon for the community. Lincoln's biography reads almost too neatly as a literal enactment of this ancient story. Born in a one-room log cabin on the Kentucky frontier, moved by his improvident father first to Indiana and then to Illinois, he educated himself by firelight, reading the Bible, Aesop's Fables, Pilgrim's Progress, and the statutes of Indiana and Kentucky. He split rails, piloted flatboats down the Mississippi to New Orleans (where he first saw the slave auction blocks that seared his soul), and taught himself law until he was admitted to the bar. Every stage of the Hero's journey is there, visible to anyone who cares to look.

But the Hero does not operate alone. He is sustained by other archetypal energies. In Lincoln we see the Wise Old Man, the Sage who sees further than others and speaks truth that is both uncomfortable and healing. The Lincoln-Douglas debates of 1858 are a perfect illustration. Stephen Douglas, known as the Little Giant, was a master of popular sovereignty's slippery rhetoric; Lincoln, by contrast, spoke slowly, deliberately, almost painfully, as though drawing words up from a deep well of moral intuition. When he said that the nation could not endure "permanently half slave and half free," he was not predicting a merely political outcome; he was articulating a law of psychic and social wholeness. A field that tolerates such a contradiction will eventually tear itself apart, just as an organism cannot long survive with incompatible organs. The Sage sees the underlying pattern.

Then there is the Caregiver, the archetype that wants to protect, nurture, and heal. Lincoln's First Inaugural Address in March 1861 is one of the most remarkable acts of political caregiving in history. Facing seven states already in rebellion and an eighth on the verge, he did not threaten; he appealed to "the better angels of our nature." He reminded the South that the government would hold its forts and collect its duties, but that he had no intention of interfering with slavery where it already existed. He ended with a plea: "We are not enemies, but friends. We must not be enemies." That is the voice of the Caregiver trying to hold the national family together even as it fractures.

The Everyman archetype is equally vivid. Americans loved to claim Lincoln as "one of us," and there is truth in the claim. He told earthy stories, cracked jokes at his own expense, wore ill-fitting clothes, and never lost his frontier drawl. When he ran for president in 1860, his supporters paraded actual fence rails he had split thirty years earlier and sang songs about "Honest Abe, the Rail Splitter." This same Everyman signed the Homestead Act, giving 160 acres of public land to any citizen willing to farm it, and he welcomed ordinary people into the White House until his secretaries had to limit visiting hours. The Everyman archetype connects the leader to the lived experience of the people; in Lincoln it was never a pose.

Finally, and perhaps most strikingly, there is the Ruler – the King archetype in Jung's terminology. This is perhaps the most startling aspect of Lincoln's career from the perspective of morphic resonance. We are accustomed to thinking of kings as hereditary figures, crowned by divine right, and surrounded by courtiers and ceremony. Lincoln was elected with barely forty percent of the popular vote in a four-way race, took office amid secession and threats of assassination, and governed a republic at war with itself. Yet his achievements in office were incredible.

In the span of just four turbulent years he:

- preserved the Union by force of arms,
- issued the Emancipation Proclamation, freeing 3.5 million human beings from slavery by executive decree,
- signed the Pacific Railway Act that linked the Pacific and Atlantic coasts of the United States and bound the continent together,
- initiated the negotiations that led to the purchase of Alaska (completed under Johnson but begun under Lincoln),
- established the first national banking system, the Department of Agriculture, and the land-grant college system through the Morrill Act.

No hereditary monarch in Europe ever achieved so much territorial, institutional, and moral transformation in so short a time. Lincoln exercised power with a kingly amplitude, yet always within the framework of republican law. He suspended habeas corpus when he believed the life of the nation required it, but he submitted the question to Congress at the first opportunity. He was, in the profoundest sense, an elected monarch – a paradox that only America could produce, and one that the morphic field of the young Republic seems to have been awaiting.

Here we come to the heart of the matter: morphic resonance. If the theory of morphic resonance is correct, the patterns of activity established by past members of a species (or a culture, or a political party) create fields that influence subsequent similar systems. The more often a pattern is repeated, the stronger the field becomes, and the more easily later individuals or groups tune in to it. Lincoln's extraordinary concentration of archetypal energies – Hero, Sage, Caregiver, Everyman, Ruler – created an exceptionally strong morphic field. Every time a Republican convention meets under red, white, and blue bunting; every time a candidate quotes the Gettysburg Address or invokes "the last best hope of earth;" every time a president speaks of "malice toward none, and charity for all," the field is reactivated. The rituals of party nomination, the rhetoric of liberty and union, the very architecture of the Lincoln Memorial – all serve as tuning devices that bring later generations into resonance with the original pattern.

We can see the effects down the decades. Theodore Roosevelt's Square Deal and his imperial ventures in Panama resonate with Lincoln's blend of Caregiver and Ruler. Eisenhower's interstate highway system echoes the

Pacific Railway. Reagan's rhetoric of freedom and his invocation of Lincoln at the 1980 and 1984 Republican conventions were explicit attempts to tune the party back to the morphic resonance of Lincoln. Even today, when Republican candidates speak of "the party of Lincoln," they are not merely indulging in ancestor worship; they are, whether they know it or not, participating in a morphic field that began to form in the oak groves of Jackson, Michigan, in 1854 and reached its first great peak in the White House from 1861 to 1865.

Lincoln's assassination on Good Friday in 1865 only strengthened his contribution to the American morphic field. Martyrdom is one of the most potent ways an archetypal pattern stabilizes itself. The Hero's death and apotheosis fix his image in the collective psyche with supernatural clarity. Lincoln's body lying in state beneath the unfinished dome of the Capitol, the funeral train retracing his journey from Springfield to Washington, the thousands who filed past the open coffin – all of these were rituals that etched the archetypal pattern deep into the American morphic field. From that moment forward, any leader who wished to claim the mantle of American purpose has had to stand, consciously or unconsciously, under the mantle of Lincoln.

The Republican Party itself can be understood as a kind of political species whose genome is forever attuned to Lincoln's archetypes. Just as organisms in nature rapidly diversify into hundreds of species by tuning to slightly different ecological frequencies, so the Republican Party has repeatedly reinvented itself – from the radical abolitionists of the 1850s to the progressive imperialists of the early twentieth century, from the conservative internationalists of the Cold War to the populist nationalists of our own time – yet always within the resonance of Lincoln's original constellation of archetypes. The field is flexible enough to allow adaptation, but strong enough to maintain identity across a century and a half.

This is not to say that every Republican politician consciously embodies Lincoln's virtues – far from it. Morphic fields influence, but they do not compel. Yet even the distortions are revealing. When a politician claims Lincoln's legacy while pursuing policies that seem antithetical to his deepest instincts, the dissonance itself testifies to the power of the field. One must still speak in its language, still gesture toward its images, because the American people continue to listen on that frequency.

New technologies are now reshaping the morphic fields of politics in revolutionary fashion–social media platforms that amplify emotion faster than reflection, algorithms that sort voters into echo chambers, artificial intelligences that can generate Lincoln speeches indistinguishable from the real thing. Yet the older fields are still there, deeper than the digital noise. Somewhere beneath the anger and the partisanship, the pattern that crystallised in the crisis of the 1850s and found its supreme focus in Abraham Lincoln continues to exert its influence.

If we listen carefully, we can still hear it: the slow, deliberate cadence of a man who wrestled with the angel of the nation's destiny and did not let go until he had received a blessing. The voice that spoke at Cooper Union in 1860, at Gettysburg in 1863, and on the blood-stained platform of the Second Inaugural in 1865 still resonates, not because it is preserved in books (though it is), but because it is preserved in the living morphic field of the American mind. So long as that field endures, the possibility remains that another leader – or perhaps a whole generation of citizens – will attune themselves to its frequency and discover anew what it means to be the Hero, the Sage, the Caregiver, the Everyman, and the elected Ruler of a people who aspire, against all the odds, to form a more perfect union.

XIX

Morphic Resonance and Political Speciation: A Natural History of Ideological Evolution in Contemporary Europe

The living world does not evolve by slow, continuous transformation alone. From time to time it leaps. A small population, isolated by geography, behaviour, or sheer chance, begins to accumulate differences that are at first almost imperceptible. Then, quite suddenly, the gulf becomes unbridgeable. The daughter population can no longer interbreed with the parent; the songs, scents, or courtship dances that once drew them together now elicit indifference or hostility. A new species has come into being. The same laws of formative causation that govern the appearance of a new finch on a remote archipelago or a new cichlid fish in an African lake also govern the appearance of a new political movement in the collective mind of a nation. The process is not metaphorical. It is homologous. Political speciation follows the same sequence of resonance-governed stages that we have used in this book to describe biological morphogenesis: a seeded perturbation, fading parental resonance, self-reinforcing divergence, threshold crossing, reactive superposition, and final stabilization of the new political movement in its own separate morphic field.

Across Europe in the early twenty-first century we have watched this drama unfold again and again, with a rapidity that would astonish a Darwinian biologist accustomed to timescales measured in tens of thousands of years. In little more than a decade, the continent's political ecosystem has been transformed and reassembled into forms that political observers from 2005 would have scarcely recognized. Podemos, La France Insoumise, the Alternative für Deutschland, the Brexit Party/Reform UK current, the gilets jaunes constellation, Sahra Wagenknecht's BSW, the slow-motion disintegration of Italian and Spanish party systems—all are daughter political species that have crossed the speciation threshold and now live by their own habits of thought and feeling. To understand how this happened, and why reversal is impossible once the threshold is passed, we must follow the same six-phase sequence that governs the origin of a new species in nature.

1. The Seeded Perturbation: Daughter Ideology Initialization

In biology, no new species appears ex nihilo. Every daughter lineage begins as a peripheral population within the parent species, carrying virtually the entire genome of its ancestor plus a small number of crucial changes—perhaps a shifted flowering time, a new mating call, a different preference in food plants. The same is true in politics (see Equation 1). When Podemos was formally launched in January 2014, its founding manifesto was seeded almost entirely from the living morphic field of the Spanish Socialist Workers' Party (PSOE): the same red banners, the same historical references to the Second Republic and the anti-Franco resistance, the same emotional claim to represent "the people" against oligarchy. What differed was a minute but decisive perturbation: the refusal to accept any further austerity memoranda imposed by Brussels and Frankfurt. That single innovation—amplified first in the plazas of the 15-M movement and then in television studios—was enough to set an irreversible divergence in motion.

We see the identical pattern elsewhere. The Alternative für Deutschland was registered in February 2013 by economics professors who disagreed with Angela Merkel's Eurozone bail-outs. At the moment of founding, its ideological genome was indistinguishable from the liberal-conservative wing of the CDU/CSU except for that one mutation: the conviction that the introduction of the euro had been a catastrophic error. La France Insoumise began in February 2016 as a personal campaign vehicle for Jean-Luc Mélenchon, who had left the Parti Socialiste in 2008. Its initial programme was 90 per cent inherited from French socialism—secular republicanism, defence of public services, anti-imperialism—altered only by a new insistence on ecological planning and a Sixth Republic. In every case the daughter movement is born almost identical to the parent, carrying its songs, its myths, its voter reflexes, its organisational DNA. The novelty is at first no more than a new inflection, a shifted emotional valence, a refusal where the parent had learned to acquiesce. Yet that is all evolution ever requires.

2. Fading Parental Resonance: The Half-Life of the Old Guard

In biological speciation, hybrid zones can persist for thousands of generations while selection and drift slowly reduce gene flow. Eventually, however, the resonance between parent and daughter populations decays until mating attempts produce only sterile or unfit offspring. Political resonance decays far more rapidly, governed by the exponential forgetting curve familiar from cultural transmission (see Equation 2). In the first years after a split, the parental field continues to exert a strong influence. Early Podemos rallies still sang the PSOE anthem "El Pueblo Unido"; early AfD parliamentarians still used Christian-Democratic phrases like "soziale Marktwirtschaft"; Mélenchon's 2017 campaign borrowed wholesale from the iconography of Mitterrand's 1981 victory.

Yet the half-life of this resonance is astonishingly short—often no more than three to five years. Each new cohort of activists arrives knowing the parent only as a betrayer rather than a living tradition. By 2019, few Podemos voters under thirty had ever cast a ballot for the PSOE in its prime; by 2021, AfD members in Saxony regarded the CDU as a foreign party; by 2023, La France Insoumise supporters spoke of the Parti Socialiste in the same tone that French communists once reserved for the SFIO. The decay is accelerated by deliberate acts of differentiation: Podemos abandoned red for purple, La France Insoumise replaced the rose with the Greek phi, the AfD exchanged the CDU's black for a new cornflower blue. Distance in political-stance-space and distance in time conspire to silence the parental echo until one day the two groups discover they no longer share a common emotional vocabulary.

3. Self-Reinforcement and Stochastic Innovation: The Evolution of the Daughter Field

Once the parental echo begins to fade, the daughter lineage comes under the governance of its own morphic field. In biology this is the phase of rapid adaptive radiation as the new species explores possibilities that were suppressed in the ancestral niche. In politics it is the explosive creativity of the first legislatures, the first mass campaigns, the first viral memes (see Equation 3). The self-resonance term dominates wherever activist density is high: in the círculos of Podemos that met weekly in every neighbourhood, in the Discord servers and Telegram channels that coordinated La France Insoumise during the 2022 legislative campaign, in the alternative media ecosystem (Junge Freiheit, Tichys Einblick, Compact) that sustained the AfD's eastward pivot after 2015.

These new habitats function exactly like the new islands or lakes that receive colonising finches or cichlids: the moment the restraining influence of the parent is lifted, variations that were once marginal become central. What was whispered in private Weinkeller or faculty lounges becomes the new orthodoxy. At the same time, stochastic innovations—analogous to the random mutations that fuel biological evolution—continue to arrive: the Catalan procés of 2017 that pushed Podemos toward explicit plurinationalism, the gilets jaunes revolt of 2018–19 that gave Mélenchon his defining image as the only national leader to march with them, the Covid lockdown protests that fused libertarian and identitarian currents within the AfD. These shocks are unpredictable in detail but inevitable in their cumulative effect: in mathematical terms, they prevent the daughter lineage from simply oscillating back toward the parent's statistical mean.

4. Crossing the Speciation Threshold

Biologists recognise speciation as complete when gene flow between populations falls effectively to zero. In politics the equivalent moment arrives when the overlap between morphic fields drops below a critical threshold (see Equation 4). Empirically this occurs at roughly 0.3–0.4 normalised similarity—precisely the range observed in lasting European splits. When Podemos secured 20 per cent of the vote in the 2015 and 2016 general elections while refusing any investiture of the PSOE without a referendum on coalition terms, the speciation was complete. When La France Insoumise emerged from the 2022 legislative elections with 75 deputies who sat as a separate group and refused any formal arrangements with the remnants of French socialism, the threshold had been crossed. When the AfD overtook the CDU in every eastern Landtag in 2024 and 2025, the two parties ceased to be variants of German conservatism and became distinct species occupying different emotional and psychic territories.

The signs are unmistakable: separate European Parliament groups, separate summer universities, separate donor networks, and separate youth organizations that no longer recognise the parent party's authority. Journalists, pollsters, and eventually voters update their cognitive maps. What was once "the radical wing" becomes a different party. Attempts at hybridization—Sánchez's coalition with Yolanda Díaz, the German traffic-light coalition's flirtation with Die Linke—produce only the political equivalent of sterile mules: governments that limp along for a season and then collapse, leaving both parent and daughter weakened but still distant and distinct.

5. Reactive Superposition

No daughter species defines itself solely by its novelties. Every new species carries its ancestor within it as a negated image (see Equation 5). The Galápagos finches sing modified versions of the mainland song; the daughter ideology speaks a modified version of the parental tongue, only with certain words inverted. Podemos kept the PSOE's anti-francoist memory while making the Transition itself the original sin. La France Insoumise kept the Parti Socialiste's secular republicanism while denouncing its European treaties as new versions of Versailles, this time imposed on France. The AfD kept the CDU's rhetoric of Heimat and Leistung while turning its welcoming culture on its head.

This reactive superposition is the source of the extraordinary emotional temperature of every successful breakaway movement. The daughter needs the parent both as foil and as fund of moral capital. The fiercer the denunciation of betrayal, the more passionately the inherited symbols are brandished—often symbols the parent itself had begun to neglect. The purple tide in Spain waves redder flags than the PSOE had dared since 1982; Mélenchon's tricolour is more fiercely republican than any official ceremony of the Fifth Republic; the AfD's nostalgia for the old Bundesrepublik is more intense than that of the CDU leaders who actually helped to govern it.

6. Stabilization in Separate Niches: The New Ecological Equilibrium

Finally the ideological landscape partitions into stable basins of attraction (see Equation 6). The daughter species settles into its own morphic cloud, sustained by the habitual activity of its militants and repelled from the parent by the memory of past betrayal. Podemos occupies the niche of indignant anti-austerity republicanism; La France Insoumise the niche of popular-ecological Sixth-Republic sovereignty; the AfD the niche of eastern German national-conservatism; Reform UK the niche of English working-class Euroscepticism. These niches are as real as the different feeding depths of Darwin's finches or the different spawning grounds of Lake Victoria cichlids. They are maintained not by geographical barriers but by resonance barriers—by habits of thought and feeling that now pull in opposite directions.

Archipelagos of Morphic Fields

The European political ecosystem of November 2025 is therefore no longer a single left–right continuum with moderate and extreme wings. It is an archipelago of morphic fields, some ancient and diminished (the Parti Socialiste reduced to a few dozen deputies, the Spanish PSOE governing only by grace of its former daughter, the Italian Democratic Party a pale shadow), some newly stabilized and radiating into sub-species of their own (the Sahra Wagenknecht split from Die Linke, the multiple fragments of the Five Star Movement, the slow coalescence of a post-Macron centre that is neither Gaullist nor social-liberal in the old sense). Parent species wither or survive in diminished form; daughter species proliferate and begin their own cycles of tension and potential fission.

This is the profound meaning of political speciation under morphic resonance. New forms do not arise because abstract reasoning discovers a better political program. They arise because the accumulated habits of a human group reach a crisis of self-contradiction, a small perturbation is amplified by passionate enactment in new social conditions, parental influence fades, random shocks push the trajectory of change further, a threshold is crossed, identity crystallizes through a combination of inheritance and reaction, and finally two self-organizing fields settle into separate niches where they will evolve along independent paths for as long as their respective habits of mind and feeling are repeated by succeeding generations.

The process is as lawful as anything in biology, and far faster. Where the origin of a new bird species may require ten thousand generations, the origin of a new political species in the age of digital resonance requires only one or two. The Europe of 2025 is living proof of this concept. Its political bestiary is richer, stranger, and more dynamically unstable than at any time since 1945, not because human nature has changed, but because the laws of formative causation operate with merciless fidelity whenever the tensions within an old morphic field become unbearable. New species emerge that simultaneously carry the past within them, reacting against it, and slowly forgetting it as their own habits become nature. And once the threshold of speciation is crossed, there is no path of return—only the long, ramifying future of an ideological tree whose newest branches continue to diverge under the same ancient mathematical laws that brought the first cell, the first animal, and the first human society into being.

The equations that describe this process are not ornaments. They are attempts to formalize, in the only language precise enough to capture resonance across the three dimensions of time, what the plazas of Madrid in 2011, the French roundabouts of 2018, and the Parisian polling stations of 2024 already demonstrated: something new had been born that could never again be unborn. The political world, like the biological world, evolves by speciation. And Europe, more than any other continent in our time, has become the Galápagos Islands of the political mind.

Mathematical Appendix for Political Speciation

(Adapted from the morphic-resonance model of biological speciation to the emergence of new political ideologies/parties/movements from a parent ideological field)

1. Daughter Ideology Initialization (Splitting Event)

When a faction breaks away (e.g., a new populist wing splits from a traditional conservative party, or a progressive caucus splits from a center-left party), its initial ideological field is seeded from the parent field plus a small mutational perturbation representing the novel positions that triggered the split.

$$I_D^i(t_{\text{split}}) = I_P^i(t_{\text{split}}) + \varepsilon\, \Delta I_{\text{trigger}}^i(t_{\text{split}})$$

- I_D^i: morphic ideological field of the daughter faction (vector in ideological space; components may represent positions on economic redistribution, social liberalism, nationalism, etc.)
- I_P^i: parent ideological field at the moment of split
- $\varepsilon \ll 1$: small mutation amplitude (size of the triggering grievance or charismatic innovation)
- $\Delta I_{\text{trigger}}^i$: direction of the novel ideological perturbation (e.g., anti-establishment, eco-socialist, libertarian-populist vector)

2. Parental Resonance Kernel (Fading Influence of the Old Guard)

The influence of the original party/ideology on the daughter faction decays with both temporal distance and ideological distance.

$$K_{D\leftarrow P}^i = G_{\text{div}}(ds_{PD}^2)\, \delta_j^i\, e^{-\lambda(t-t_{\text{split}})}$$

- $G_{\text{div}}(ds_{PD}^2)$: Gaussian function of ideological distance in "stance space" $ds_{PD}^2 = \|I_D - I_P\|^2$
- λ: rate of cultural forgetting / generational turnover
- This kernel multiplies parent-field contributions; as the daughter moves farther in stance space or time passes, parent resonance fades exponentially.

3. Daughter Ideological Field Evolution

The daughter ideology evolves under three forces analogous to the biological case:

- Self-reinforcement within the new faction
- Residual echo from the parent ideology
- Stochastic innovation (new leaders, viral memes, external shocks)

$$\frac{\partial I_D^i}{\partial t} = \kappa_D K_{Dj}^i \phi_D^j + \kappa_{D\leftarrow P} K_{D\leftarrow P_j}^i \phi_P^j + \eta \xi^i(t)$$

- κ_D: self-resonance strength of the daughter faction (higher with strong online communities, charismatic leaders, or youth mobilization)
- $\kappa_{D\leftarrow P}$: cross-resonance strength from parent (often negative in practice as the daughter defines itself in opposition)
- ϕ^j: local activist density field (where ideas are actually debated and enacted)
- $\eta \xi^i(t)$: Gaussian white noise representing unpredictable ideological innovations (new slogans, scandals, geopolitical events)

4. Speciation Threshold (Declaration of Distinct Political Identity)

A new political entity is recognized as an autonomous "species" (separate party, movement, or lasting faction) when the normalized overlap of its morphic ideological field with the parent falls below a critical threshold θ_c (empirically ~0.3–0.4 for lasting splits).

$$\mathcal{O}_{DP}(t) = \frac{\int I_D^i I_P^i \, d^3t}{\sqrt{\int I_D^j I_D^j \, d^3t \cdot \int I_P^k I_P^k \, d^3t}} < \theta_c$$

- The integral is taken over Kletetschka's three-dimensional time manifold (past ideological states continue to exert influence non-locally).
- When overlap drops below θ_c, journalists, voters, and elites begin treating the daughter as a distinct entity; fundraising, media coverage, and ballot access follow.

5. Ideological Form Bifurcation (Post-Split Identity Crystallization)

After the split, members of the daughter faction psychologically and rhetorically construct their new identity as a weighted superposition of:

- Inherited tradition from the parent
- Radical novelty
- Reactive entanglement with the parent (defining oneself *against* the old guard)

$$\psi_D^i = \sqrt{1-p}\,\phi_P^i + \sqrt{p}\,\phi_{\text{novel}}^i + \sqrt{p(1-p)}\,\Phi_{\text{anti-P}}^{ij}\phi_P^j$$

- $p \in [0,1]$: novelty parameter (low p = moderate reformist split; high p = revolutionary or purity-spiral split)
- $\Phi_{\text{anti-P}}^{ij}$: antisymmetric tensor that flips the sign of parent positions the daughter most rejects (e.g., pro-immigration parent → anti-immigration daughter, or vice versa)
- This equation produces the characteristic love–hate relationship: the daughter inherits organizational know-how and some voters from the parent while simultaneously defining itself in opposition.

6. Morphic Cloud Partition (Stabilization of Separate Ideological Niches)

Once speciation is complete, the daughter ideology carves out its own stable region in ideological space via attractive self-resonance within its base and repulsive gradients away from the parent cloud.

$$\frac{D^2\gamma_D^k}{d\lambda^2} = \kappa_D K^k_{Dl}\,\delta\phi^l_D - \kappa_{D\leftarrow P}\,\nabla^k \mathcal{O}_{DP}$$

- First term: attractive potential pulling activists and rhetoric toward the daughter's emerging consensus
- Second term: repulsive gradient pushing the daughter cloud away from maximal overlap with the parent (avoids re-absorption)
- Solutions are stable fixed points corresponding to distinct ideological niches (e.g., MAGA vs establishment GOP, Corbynite Labour vs New Labour, DSA vs mainstream Democrats, AfD vs CDU, etc.).

These six sets of equations provide a direct mathematical translation of the original biological speciation model into the political domain, preserving the core ideas of:

- seeded divergence,
- fading parental resonance,
- self-reinforcing evolution with noise,
- a measurable speciation threshold,
- psychological superposition of inheritance and reaction,
- and final stabilization into separate morphic clouds in the ideological landscape.

They can be numerically simulated (e.g., in Python with NumPy/SciPy) to study real historical splits (Brexit Party from UK Conservatives, Podemos from PSOE, Tea Party from GOP, La France Insoumise from PS, etc.) and to forecast when current intra-party factions are likely to undergo full political speciation.

XX

A Mind With Two Souls

The mind is not confined within the skull. It extends across time and space through morphic resonance, the process by which similar patterns of activity in the past influence similar patterns in the present. Past minds tune present ones, not through any material inheritance but through fields of habitual organisation that persist as long as the patterns themselves are repeated somewhere, by someone. Each person is therefore surrounded and interpenetrated by a hierarchy of fields housed in the three dimensional architecture of time: those of immediate family, of language, of culture, and, deeper still, of ancestral lineages. These fields do not fade proportionally with generations; they remain accessible, sometimes faintly, sometimes with startling clarity, depending on the degree of similarity between a present life and those in the past that gave the field its original form.

Let us imagine a person called Sofia Alexeyevna Alighieri. She is thirty-seven years old. She lives in a small brick house on the edge of a college town in western Massachusetts, where the winters are long enough to remind her of places she has never seen. She teaches comparative literature, publishes poems that refuse to settle into any single tradition, and often awakes with the sensation that two distinct centuries are breathing through her at once.

Through her father she descends, however obliquely, from Catherine the Great. The line begins with one of the Empress's unacknowledged children by Grigory Potemkin, passes through Baltic-German nobility, through White Russian exiles who crossed frozen seas with nothing but icons and pride, and ends with a mild American professor of Slavic languages who married a local Italian American woman in 1976. Through her mother she descends from Dante Alighieri himself. The Florentine branch of the family clung to its name and its tower for five centuries after the poet's exile, survived plague and taxation, and finally scattered in the twentieth century; one penniless countess married an Italian-American engineer from New Jersey, and their granddaughter is Sofia.

Two ancestral fields of almost unimaginable dissimilarity now intersect in the same extended mind.

The Catherinian field is vast, theatrical, and unapologetically imperial. It is tuned to horizons that most human beings never learn to see: the sense that a single will can redraw maps, build cities overnight, and still have energy left for philosophical correspondence and passionate love affairs before luncheon. It carries the habit of treating Europe as a private stage, of speaking French with Voltaire one week and ordering the annexation of Crimea the next. It is a field of sensuality without shame, of Enlightenment reason wielded like a sceptre, of a woman who could laugh at her own ambition even while pursuing it to the edges of the known world.

The Dantean field is narrow, vertical, and incandescent. It is organised around a vision of absolute justice in which every betrayal, every kindness, every half-truth is weighed on cosmic scales. It is a field of exile, of language honed until it can slice the soul in two, of a love so fierce that it demands the remaking of the entire universe. It is the habit of seeing the city street as a prelude to judgment, of hearing in every human voice either the music of paradise or the whisper of damnation.

Sofia first noticed the Russian field in childhood. At eight she began dreaming in a language no one in her house spoke. She would wake with the taste of kvas and the sound of sleigh bells, speaking phrases that belonged to the court of 1780 rather than to any living Russian. Years later, when she finally studied the language, she discovered that the sentences she had murmured in her sleep were drawn almost verbatim from Catherine's private letters. No book in her parents' house contained them; no recording had ever played them. The field had simply found a receiving set sufficiently similar to its original source.

The Dantean field announced itself more dramatically. At twenty-three, while copying a fourteenth-century manuscript in the Laurentian Library, she was overcome by a light so intense that she had to sit on the stone floor. For three days she saw modern Florence overlaid by a city of fire and gold, its streets filled with figures who turned to her with recognition and accusation in the same glance. She wrote hundreds of lines of terza rima in a single night, in a Tuscan so archaic that the Florentine scholars who later examined it could date it, almost to the decade, to the early 1300s. When she recovered, she understood that she had been tuned, abruptly and without warning, to the frequency of her exiled ancestor.

Since then the two fields have lived together inside her, sometimes in harmony, more often in creative friction.

On mornings when the Russian field predominates she feels the continent inside her chest. She walks across campus with an unexplained sense of territorial possession, as though the quadrangles and libraries were merely temporary encampments on a plain that stretches to the Pacific. Her gestures grow wider, her laughter more operatic. She will catch herself planning imaginary universities in the steppes or writing memos with the crisp decisiveness of an imperial rescript. Students leave her lectures slightly dazed, as if they have been addressed by someone who expects nations to obey.

On mornings when the Dantean field rises she cannot write a grocery list without sensing the gravity of judgment in every word. Language becomes a court of last appeal; the wrong preposition feels like perjury. She walks beneath the autumn trees hearing the precise circle of Hell to which each falling leaf is consigned. Love, on those days, is almost unbearable: desire feels like a betrayal of the celestial order, yet the memory of Beatrice (never read, only remembered) makes abstinence equally impossible.

Most days the fields do not alternate; they interfere.

When she falls in love (which happens rarely and against her better judgment) the Russian field wants to annex the beloved outright, to build palaces for him, to issue manifestos in his name, to make him co-regent of an empire that exists only in her mind. The Dantean field, meanwhile, wants to weigh his soul on scales of starlight, to know whether this attachment will lead upward toward the Empyrean or downward through the ice of Caina. The poor man usually finds himself confronted by a woman who oscillates, sometimes in the same evening, between Catherine's magnanimity and Dante's intransigence. Few survive the contradiction.

Her poetry is the place where the two fields achieve their strangest and most fertile reconciliation. A single sonnet may begin with the snows of the Russian steppe and end in the frozen silence of Satan's wings; another will move from the glittering chandeliers of Tsarskoye Selo to the dark wood where Dante first lost the straight way. Reviewers speak of "anachronistic splendour" and "ontological vertigo." They do not know that the poems are not inventions but transcriptions of an inner weather system as old as time.

She has learned certain techniques of navigation. When the imperial field threatens to overwhelm her with grandiosity she recites Dante's line about the love that moves the sun and the other stars, and the perspective narrows to a single human heart. When the Dantean field begins to condemn the world wholesale she remembers Catherine's delighted discovery that human beings are, after all, rather charming creatures if one gives them palaces and occasional wars, and laughter returns. The two medicines are bitter to one another, yet each prevents the other from becoming fatal.

Some evenings she stands at her kitchen window watching the New England dark come down and feels the entire history of Europe leaning against the glass, waiting to be let in. She understands then that she is not the end of two declining aristocratic lines but the place where two immense habits of being have chosen to meet again after seven centuries apart. The fields did not seek her out because she is particularly virtuous or particularly wicked; they sought her because her neural organisation, her emotional temperament, her very syntax, happen to resonate with both frequencies at once. She is a kind of living interferometer, an instrument designed (without anyone's intention) to measure the distance between empire and exile, between sensuality and salvation, between the eighteenth century's confident morning and the fourteenth century's implacable noon.

And because morphic fields grow stronger wherever their patterns are repeated, every poem she writes, every lecture she gives, every act of forgiveness or ambition she performs in her small American life adds, in however microscopic a way, to the combined field that will be inherited by whoever comes after her with a similar mixture of blood and memory. The resonance may even move forward as well as backward; Catherine and Dante themselves may have been subtly altered by what their remote descendant feels and thinks under a sky they never imagined.

This is the true meaning of ancestry in a universe of morphic resonance: not a fading echo but an ongoing conversation in which the living are constantly influenced by the habits of thought and action in the past. Sofia Alexeyevna Alighieri, eating pelmeni and drinking Chianti in Massachusetts while two empires and one cosmos argue inside her skull, is simply the present location of that conversation. Tomorrow the voices may find another host. For now they have chosen her, and she has learned to live with the splendid, terrifying amplitude of being addressed by history itself.

XXI

Morphic Resonance and the Memory of a Vanished World

A civilization does not die all at once. Walls crumble, cities drown, bodies decay, yet the habits that gave that civilizations its form persist in the memory of nature itself. Morphic fields are fields of memory, embedded in the three dimensional manifold of time, sustained by repetition, strengthened by collective performance, and governed by the same laws that organize the growth of crystals or the flocking of birds. When a whole world of ritual, measurement, architecture, astronomy, agriculture, and seafaring is suddenly shattered, the field does not vanish; it becomes orphaned. It lingers, attenuated but intact, like a vast chord struck twelve thousand years ago that has never quite ceased to sound.

This essay explores the hypothesis, advanced with considerable evidence by Graham Hancock in America Before (2019), that an advanced, globally influential civilization flourished during the latter part of the last Ice Age, reached its height roughly 12,900 years ago, and was largely extinguished soon thereafter by the multiple impacts of a fragmenting comet or meteorites from the Taurid meteor cloud. The comet fragments or meteorites would have struck the then massive North American and European ice sheets as well as lands not covered by the ice, causing massive melting and starting immense fires over vast areas. The immediate impact on the global climate would have resembled that of a major nuclear war, with huge clouds of smoke, ash, and dust blocking the light of the sun, much like a "nuclear winter." . The impacts seem to have precipitated the Younger Dryas period and caused meltwater pulses that raised sea levels by approximately 120 meters in stages. For over a thousand years, and plunged surviving human populations into a dark age of trauma and forgetting.

Yet the morphic fields of that lost civilisation did not perish. They decayed, certainly, but because they had been built by tens or perhaps hundreds of millions of people performing the same patterns countless times, their residual amplitude remained greater than any new fields the traumatized remnants could generate from scratch. The result was a twelve-millennia-long echo that manifested wherever population density and cultural receptivity rose above certain thresholds. Göbekli Tepe, the Amazonian geoglyphs, the sudden domestications of wheat and maize, the shared precessional astronomy of Egypt and Mesoamerica, the bearded culture-bearers of American myth, and even a 6,000-year-old Armenian wine jar that records a solar eclipse and the Saros cycle—all can be interpreted as partial re-manifestations of an ancestral morphic cloud that never entirely dispersed.

The Cataclysm and the Orphaned Field

The evidence for the comet impact itself has grown robust since the 2007 publication of the Younger Dryas Impact Hypothesis. Nanodiamonds, microspherules, platinum spikes, and widespread black mats of charred material appear synchronously across four continents at the Younger Dryas boundary (~12,800 years before present). Meltwater Pulse 1A, one of the fastest sea-level rises known in the geological record, followed almost immediately. Coastal plains that had been dry land for tens of thousands of years vanished beneath rising waters. The centers of population, ports, observatories, and libraries of any advanced Ice Age civilization would have lain precisely in the zones that are now continental shelves. The absence of obvious ruins is therefore not evidence of absence; it is evidence of drowning.

In morphic terms the mathematics are straightforward. Let F_0 be the amplitude of the pre-catastrophe global field, built by a population P_0 over many generations. Field strength scales roughly with the cube of the number of tuned nervous systems (the non-linear collective term in Equation 3 of the Mathematical Annex). If $P(t)/P_0$ collapses from 1 to 0.01 or less within a few decades, the orphaned field $F(t)$ falls to approximately $10^{-6} F_0$ almost at once. Natural temporal decay adds another factor of $e^\wedge(-\lambda t)$ where λ is small for deeply habituated patterns. For the first few thousand years the orphaned field is therefore a million times weaker than it was, yet still roughly a thousand times stronger than any new field the survivors can create from improvised behaviour. It is a ghost, but a very loud ghost.

The First Punctuation: Göbekli Tepe and the Re-awakening of Sacred Architecture

By 9600 BC the Younger Dryas had ended, sea levels were stabilising at something close to modern heights, and human population was climbing again in favoured refugia. At that moment the archaeological record registers a shock: bands of hunter-gatherers in south-east Anatolia begin quarrying, shaping, transporting, and erecting T-shaped limestone pillars weighing up to twenty tons. They arrange them in perfect circles, carve sophisticated reliefs of animals that were already becoming rare, and decorate the pillars with abstract symbols that will not be seen again for millennia. The oldest enclosures at Göbekli Tepe pre-date settled agriculture, pottery, and domestic animals in the region by centuries.

Orthodox explanation falters. How do pre-agricultural foragers mobilise the labour of hundreds for projects that yield no obvious caloric return? In morphic terms the answer is clear: the builders were re-enacting a habit of sacred circular architecture that had been performed millions of times before the comet struck. The orphaned field of megalithic temple-building, though attenuated, still exerted sufficient attractive force to organise hands that had never been taught the craft. The same resonance appears contemporaneously at Karahan Tepe, Taş Tepeler, and a dozen other sites now emerging across the region. The punctuation is not local; it is continental, exactly as the non-local term in Equation 6 predicts.

The Amazon and the Americas: Echoes across a Drowned Continent

Across the Atlantic a parallel story unfolds, though the evidence has only recently become visible. LiDAR surveys conducted since 2015 have stripped away the forest canopy of Amazonia to reveal an astonishing lattice of geoglyphs, causeways, raised fields, and monumental earthworks covering an area larger than Great Britain. Orthodoxy had insisted that Amazonia before 500 AD could support only small, dispersed populations. The new data demand large, co-ordinated societies by at least 1500 BC, and probably centuries or millennia earlier. Ceramic dates are pushing back toward the early Holocene, and some earthworks now appear to pre-date the Younger Dryas itself.

Hancock's argument is that many of these structures belong to the lost Ice Age civilisation or to its immediate descendants. Submerged structures off the

Yucatán, the Bimini Road, and the Yonaguni formation hint at what has been lost beneath the waves. The sudden appearance of maize domestication in Mesoamerica, bottle-gourd cultivation in the Andes, and manioc terra preta agriculture in Amazonia—all within centuries of each other and thousands of kilometres apart—follows the same punctuated pattern seen in the Old World. Morphic resonance propagates instantly across distance when the field is strong enough; oceans are no barrier to a habit that once spanned the globe.

The Bearded Culture-Bearers and the Transmission of Knowledge

Ancient American mythologies preserve what appear to be historical memories of the survivors. The Inca spoke of Viracocha, a white-skinned, bearded figure who emerged from Lake Titicaca after a terrible deluge, taught agriculture, architecture, and law, then departed across the Pacific. The Maya and Aztecs remembered Quetzalcoatl/Kukulkan in almost identical terms: a bearded bringer of the calendar, writing, and maize who sailed away promising return. The Chimu, the Chachapoyas, the Aymara—all tell variants of the same story. These are not late inventions; they are embedded in the earliest strata of oral tradition and iconography.

Mainstream archaeology treats such myths as late syncretism or post-Conquest fantasy. Yet the shared motif of bearded, light-skinned teachers arriving by sea after a flood is too specific and too widespread to dismiss. In a morphic interpretation the culture-bearers were real: survivors from the drowned coastal cities who carried with them the residual field of astronomy, pyramid-building, and intensive agriculture. Where the field encountered receptive populations—traumatised hunter-gatherers whose own fields were weak—the ancient habits re-crystallised with startling speed. The result was the sudden appearance of Olmec heads, Chavín de Huántar, Caral-Supe, and the Amazonian earthworks: all "too early," all too sophisticated, all echoing forms known from Göbekli Tepe, Malta (the island), and Gunung Padang on the other side of the planet.

The Cosmic Vessel of Areni-1: An Atavistic Masterpiece

In 2007 archaeologists working in the Areni-1 cave complex in Armenia uncovered a large red-clay karas dated by a dozen radiocarbon samples to approximately 4100–3900 BC. The vessel is not a humble wine jar. Its entire surface is covered with black-pigment drawings that interweave a mythological combat between horned sheep (light, order) and serpent-dragons (darkness, chaos) with an astonishingly precise astronomical notation. At the centre stands a black sun surrounded by a radiant corona and daytime-visible stars: an unmistakable representation of a total solar eclipse. Computer reconstructions of the ancient sky above the cave confirm that a total eclipse did indeed occur on 11 April 3958 BC, visible from that exact location.

Around the eclipse imagery appear symbols that scholars now recognise as an early encoding of the 223-month Saros cycle and the 19-year Metonic cycle—mathematical tools for eclipse prediction that supposedly were not discovered until Babylonian times two thousand years later. Adjacent symbols delineate a 365-day solar year divided into twelve months with intercalation markers. The vessel was found in a ritual context beside the oldest known shoe, the oldest known winery, and vast quantities of grape pips and stems.

No Chalcolithic village should possess such knowledge. The Saros cycle requires systematic observations over at least two centuries; the Metonic cycle demands even longer. Yet here they are, painted on a wine vessel six thousand years old. Morphic resonance offers the most parsimonious explanation: a potter or priest-artisan, working within the reviving field of viniculture that had crystallised in the same cave, momentarily re-entered the full amplitude of the pre-catastrophe astronomical field. The knowledge was not laboriously re-invented; it was remembered by hands that had never been taught. The Cosmic Vessel is an atavism, a flare of the orphaned field burning bright enough to illuminate, for one brief moment, the depth of what was lost.

Why the Recovery Remains Partial

Three constraints continue to inhibit full re-ignition of the ancient field:

1. Population bottleneck and sub-field extinction. Many specialised resonances—those requiring rare metals, oceanic navigation on an industrial scale, or populations in the tens of millions—fell below critical mass and were permanently lost.

2. Loss of material templates. Resonance can guide the hand to carve a T-pillar or brew wine, but it cannot conjure copper out of drowned mines or recreate seed banks washed away by meltwater pulses.

3. Trauma-induced cultural inhibition. The universal myth of a golden age destroyed by divine wrath functions as a psychic brake. From the Greek hubris stories to the Maya katuns of destruction, humanity has carried a collective

memory of punishment for over-reaching. Full recovery is therefore disguised as "primitive" revival or blocked altogether.

The Present Moment

Today eight billion human beings, linked by instantaneous communication and augmented by artificial intelligences that perform patterns trillions of times per second, have raised the global proliferation constant $\kappa(t)$ to levels probably higher than at any time in human history. The new morphic field we are generating is drowning out the ancient echo. Yet the orphaned resonance still flickers in unexpected places: in the sudden recovery of lost agricultural techniques, in the unexplained precision of certain ancient measurements, in the haunting similarity of myths separated by oceans.

The cloud that does not die has accompanied humanity for twelve millennia. It carried the memory of star-charts across drowned continents, guided hands to raise Göbekli Tepe and the Amazonian geoglyphs, whispered the Saros cycle to an Armenian potter, and preserved the image of bearded survivors in the oral traditions of the Andes and Mesoamerica. The field is attenuated, fractured, traumatised—yet unbroken.

As population, connectivity, and courage increase, fragments of that ancient song grow louder. The question is no longer whether the lost civilisation existed; the geological, archaeological, mythological, and morphic evidence converge too powerfully for serious doubt. The question is whether humanity, having spent twelve thousand years mistaking echoes for original compositions, now possesses the maturity to allow the full chord to sound once more—or whether the ancient fear of hubris will again silence it.

Nature, after all, is a realm of habit. And habits, once formed by sufficient repetition across sufficient minds, acquire a kind of immortality. The morphic cloud of a vanished world still floats above the Earth, waiting patiently for the day when there are again enough hands, enough eyes, and enough hearts unafraid of their own past to let it sing at full volume.

XXII

The Great Acceleration: Why Technological Progress is Exploding

Imagine a woman two and a half million years ago, crouched on the bank of a sluggish river in what is now Ethiopia. She lifts a water-worn cobble, weighs it in her palm, and strikes it sharply against another stone. A flake falls away, its edge keen enough to cut hide. She turns the flake in the sunlight, tests it with her thumb, and nods. Something has come into the world that was not there a moment before: not merely a sharp edge, but a possibility, a habit waiting to be repeated.

That first Oldowan flake did not vanish when she died. It left behind a trace—not in genes, not in language, but in the nature of things themselves. Each time another hand, anywhere on the planet, repeated the same sequence of gestures, the same angle of blow, the same acceptance of the resulting flake, the pattern grew stronger. By the time Homo erectus was carrying fire to Java, the morphic field of the Oldowan chopper had become a presence as real as gravity. A knapper in Spain 1.8 million years later, separated from that African woman by thousands of generations and half a continent, felt the same pull toward the same spiral fracture, the same useful jagged edge. The morphic field did not need teachers. It spoke directly to the hand and the eye.

This is the great secret of nature: she is habitual. Everything that happens leaves a trace, and similar things that happen later are influenced by all similar things that have happened before. In the philosophy of Rupert Sheldrake, this influence is called morphic resonance. It is not a metaphor, but a kind of memory inherent in nature, a memory that operates through fields—morphic fields—which organize patterns of form, behavior, and, as we shall see, technology. Thanks to the theory of three-dimensional time, we have a way of describing morphic resonance mathematically.

The Acheulean hand-axe is a most astonishing demonstration of morphic resonance. For one and a half million years—six thousand human generations—millions of individuals on three continents knapped billions of these beautiful, symmetrical, tear-shaped tools. The same proportions, the same thinning of the base, the same balance in the hand recurred with a fidelity that no modern production line has ever matched. The morphic field of the Acheulean biface grew so powerful that deviation became almost impossible. A craftsperson separated from her ancestor by a million years would have recognised every detail of grip, percussion, and pattern as readily as her own reflection.

Change, when it finally came, did not arrive by slow improvement. It arrived in a leap. Somewhere, perhaps in the Rift Valley, perhaps in the Levant, a knapper prepared a core in advance, struck platforms at deliberate angles, and detached a flake whose shape was predetermined. The first perfect Levallois tortoise-core appeared fully formed (Equation 1). The deviation from the Acheulean template was small—typically five percent in measurable attributes such as edge angle, symmetry index, or basal thinning—but it was decisive. The old field lost its grip, and a new field began to resonate (see Annex, Equation 1).

Within a geological instant—centuries, not millennia—Levallois points were being produced by Neanderthals in France and by early Homo sapiens in Israel, identical in every detail, separated by four thousand kilometres and no plausible migration corridor. The new pattern had propagated not by trade or instruction, but by resonance through the newly established field (Equation 6). The same synchrony repeats throughout prehistory: Aurignacian bladelets appearing simultaneously in Bulgaria and Spain at 46,000 years ago; microblade industries exploding from Alaska to Hokkaido within three centuries; Clovis points blanketing North America from the Arctic to the Gulf in less than four hundred years. Each time the mathematics is the same: a small but irreversible mutation (Equation 1), instantaneous non-local propagation (Equation 6), followed by centuries or millennia of faithful replication while the field strengthens (Equation 3).

For a time the old and new fields overlap. The parental echo lingers, producing those beautiful hybrid industries that fill museum cases: Chatelperronian backed blades mixed with Mousterian discs; early steam engines retaining water-wheel logic; the first automobiles still shaped like carriages without horses. The strength of the parental echo decays exponentially with both calendar time and morphological distance (Equation 2). In the Paleolithic the half-life is measured in thousands of years; in the

Industrial Revolution, in decades; in the era of software, in months. When the cosine overlap between parent and daughter fields finally falls below approximately thirty percent (Equation 4), the old way no longer feels merely outdated. It feels actively wrong—clumsy, offensive to the hand, impossible to the mind. That is the precise moment when a new tradition is born in its own right.

The tree of technology grows by this branching process. Each successful speciation adds a new independent morphic field. Where once there was one dominant field, now there are two, then four, then thousands. Ten thousand years ago the planet supported perhaps a dozen major tool traditions. Five thousand years ago, with agriculture, pottery, weaving, metallurgy, and writing, the count had risen to hundreds. By 1500 CE there were thousands of distinct crafts and sciences. Today there are millions—every programming language, every biochemical pathway, every machine-learning architecture possesses its own field, strengthened every time it is successfully repeated (Equation 3).

The more fields exist, the more surfaces there are for the next mutation to land. A flint blade offers only a handful of dimensions in which to vary. A modern microprocessor offers billions of transistors; a large language model offers trillions of parameters. Complexity itself grows logarithmically with the number of fields (Equation 7), and the aggregate rate at which new fields appear is therefore proportional to N log N (Equation 5). The mathematics is merciless: the curve of technological proliferation is double-exponential. It lies almost flat for two million years, bends gently upward in the Upper Paleolithic, rises more steeply after the Neolithic, accelerates after 1700, and now ascends vertically. The same master equation (Equation 5), with only a changing prefactor $\kappa(t)$ driven by population, connectivity, and mean complexity, fits every dataset we possess: stone-tool industries, European patents 1600–2024, scientific papers 1700–present, and GitHub repositories 2008–2025.

We have entered the steepest inflection in three million years. Artificial intelligence now participates directly in the resonance. Silicon minds repeat patterns billions of times per second, explore regions of possibility no human lifetime could traverse, and broadcast successful forms instantaneously across the planet. The mutation amplitude ε, once constrained to roughly five percent by the limits of human motor control and perception, now approaches arbitrarily small values in continuous parameter space. The entanglement tensor of Equation 6 operates at global scale, correlating a breakthrough in one laboratory with latent possibilities in every other laboratory attuned to the same ancestral chorus. The proliferation constant κ has increased another thousand-fold in the past decade alone.

Consider the sudden crystallisation of large language models between 2017 and 2025. The parent field was the recurrent neural network paradigm (RNNs, LSTMs, GRUs) that had dominated sequence modelling for two decades. In that parent field the resonance was so strong that almost every researcher instinctively reached for recurrent layers and back-propagation through time. The overlap between "what a neural network should be" and "recurrence" still stood well above the thirty-percent threshold (Equation 4) as late as 2016.

Then, in June 2017, a single paper ("Attention Is All You Need") introduced the transformer. The deviation was tiny (less than seven percent in measurable architectural dimensions), yet decisive (Equation 1). Within months, independent replications and improvements appeared simultaneously at Google, OpenAI, DeepMind, Facebook AI Research, and universities around the world. No one had to be taught; the new field simply began to resonate more loudly than the old one ever had (Equation 6). The recurrent echo halved every four to six months (Equation 2). By 2020 the cosine overlap had fallen below 0.3 (Equation 4); recurrence no longer felt merely outdated, it felt actively wrong. From GPT-3 onward, each successful scaling or architectural tweak (Mixture-of-Experts, FlashAttention, Hyena, RetNet, long-context extensions) propagated globally within days, exactly as the master proliferation equation predicts when the practitioner count is in the millions, connectivity is instantaneous, and the parameter space is astronomically large (Equation 5). We are watching, in real time, the same punctuated tempo that once required a hundred thousand years in stone tools, now compressed into a single decade of silicon.

The same mathematics governs every domain where population, connectivity, and complexity are rising. Where they stagnate—regulated industries, isolated communities, monocultural paradigms—progress freezes into Paleolithic-style stasis. Nuclear power, supersonic passenger flight, and concorde-era aerospace are modern examples of fields trapped by the regulatory suppression of novelty; their morphic resonance has been artificially silenced.

Policy implications follow naturally. It is critical to foster the diversity of fields, since monocultures are inherently brittle. A single dominant paradigm—whether Levallois points or relational databases—resists speciation until a catastrophe forces a punctuation. One can preserve the fading morphic echoes of old technologies long enough to ease the transition to the new (Equation 2), but one cannot cling to them past the independence threshold (Equation 4). We can cultivate connectivity between differing producers and innovators, but without suffocating local novelty. The entanglement tensor of Equation 6 thrives on loose coupling, not central planning. Above all, we must recognize that technological evolution is not a ladder but a branching tree governed by the seven equations in the Annex.

We are not merely accelerating. We are participating in the most rapid intensification of morphic resonance the Earth has ever known. The choir that once contained just a few thousand voices for a million years now contains trillions. Each new voice strengthens the resonance of its own kind and, by adding to the total complexity of the system (Equation 7), makes the next voice more likely. The crescendo is not leading us toward some final omega point of technological perfection. It is leading us deeper into the living mystery of a universe that remembers what it has done and makes it easier to do the same thing again—until, in a moment of creative escape, it does something new.

The mathematics remains astonishingly simple. Seven equations and a handful of constants—five percent mutation (Equation 1), thirty-percent independence threshold (Equation 4), logarithmic complexity growth (Equation 7), and the master proliferation law (Equation 5)—describe the entire tempo and mode of technological evolution from the first chipped pebble 2.6 million years ago to the latest 400-billion-parameter AI model released last month. The constants are not arbitrary; they are tuned by the archaeological and historical record across four continents and numerous independent transitions.

The universe is not a dead mechanism grinding forward according to eternal Cartesian laws. It is a great organism, evolving through habits that grow ever stronger with repetition until, suddenly, one habit gives birth to another. We are inside that organism, and the morphic fields we are now strengthening with every line of code, every engineered protein, and every optimized supply chain will resonate far beyond our own lifetimes. They will shape minds that have not yet been born, on this world and perhaps on others.

The acceleration is inevitable, but its meaning is not. We still have the freedom—while we remain the dominant resonating species—to choose which fields we strengthen and which we allow to fade. The choices we make in the next few decades will determine whether the coming morphic chorus is harmonious or cacophonous, whether it serves life or devours it.

Mathematical Annex

Equation 1: Daughter Field Initialization

$$M_D^i(t_{\text{split}}) = M_P^i(t_{\text{split}}) + \epsilon\, u^i \qquad (\epsilon \approx 0.04\text{–}0.07)$$

When a truly new form is born, it appears almost complete, carrying 93–96 % of the old pattern plus one decisive twist—usually about five percent different in the things we can measure. That small but irreversible shift is enough for the new morphic field to declare its independence and begin resonating on its own.

Equation 2: Parent Echo Decay Kernel

$$K_{D\leftarrow P}(\Delta t, d_s) = \exp\left(-\frac{d_s^2}{2\sigma_s^2}\right) \exp(-\lambda \Delta t)$$

After the new form appears, the old habit lingers as a fading echo. The more different the new thing becomes, and the more time passes, the fainter the old voice grows—halving every few thousand years in prehistory, every few decades in the age of patents, every few years in software.

Equation 3: Post-Split Daughter Evolution

$$\partial_t M_D^i = \kappa_D M_D^i + \kappa_{D\leftarrow P} K_{D\leftarrow P} M_P^i + \eta \xi^i(t)$$

Once born, the new pattern is strengthened every time someone repeats it successfully. The dying echo of the old pattern continues to fade, and tiny random variations keep occurring—most harmless, some destined to become the seed of the next great leap.

Equation 4: Independence Threshold

$$\mathcal{O}_{DP}(t) = \frac{\langle M_D | M_P \rangle}{\|M_D\| \|M_P\|} < \theta_c \approx 0.30 \pm 0.05$$

A new tradition becomes fully itself when its resonance with the old one falls below about thirty percent. At that moment the ancestral way no longer feels merely out-of-date; it feels actively wrong, clumsy, impossible.

Equation 5: Master Proliferation Equation

$$\frac{dN}{dt} = \kappa(t)\, N \log N$$

The rate at which new independent morphic fields appear is jointly proportional to (i) the current number of existing fields N(t)—each can potentially spawn a daughter—and (ii) the average complexity of those fields, which grows logarithmically with their cumulative number (Equation 7). The product N log N therefore governs the birth rate of new technological paradigms.

When κ(t) is approximately constant over a given era, the solution is strictly double-exponential:

$$N(t) \propto \exp\left(e^{\kappa t + C}\right).$$

Since κ(t) itself rises slowly (and in recent centuries exponentially) owing to growing population, connectivity, experimentation speed, and now AI participation, the observed trajectory is double-exponential or faster—the familiar "hockey stick" that lies nearly flat for millions of years and then ascends almost vertically in the modern era.

Equation 6: Quantum-like Superposition at Speciation

$$|\psi_D\rangle = \sqrt{1-p}\,|M_P\rangle + \sqrt{p}\,|novel\rangle + \sqrt[4]{(1-p)p}\,\hat{\Phi} \otimes (|M_P\rangle + |novel\rangle)$$

When a new form is ready to crystallise, many minds are already attuned to the same possibility. The first successful example does not have to travel; it releases a pattern that was already waiting in the shared field.

Equation 7: Complexity Growth

$$C(t) \approx \alpha \log N(t) + \beta$$

Every new field adds a little more richness to the total repertoire of civilisation. Because of this slow logarithmic growth in average complexity, the opportunities for the next mutation keep multiplying.

Glossary of Symbols

- M^i : morphic field vector in design + space + time
- ϵ : mutation amplitude (≈5 %)
- u^i : unit direction of the decisive twist
- d_s : morphological distance
- σ_s : characteristic morphological distance scale that controls how quickly the parental echo fades as the new form diverges (the "width" of the Gaussian similarity kernel in Equation 2)
- λ : decay rate of the parental echo (temporal component)
- $\kappa(t)$: time-dependent proliferation constant
- θ_c : independence threshold (≈0.3)
- p : novelty fraction at speciation
- $\hat{\Phi}^{ij}$: entanglement tensor (non-local correlation)
- $N(t)$: number of active independent fields
- $C(t)$: mean complexity per field
- $\xi^i(t)$: ordinary small variations

The Mathematics of Morphic Resonance

1. Temporal Metric

$ds^2 = g_{ij}\,\Delta t^i\,\Delta t^j$

(multiplication between g_{ij} and Δt^i; dot product implied by Einstein summation over i, j between Δt^i and Δt^j)

This equation defines a distance measure in a multi-dimensional "time space" where each axis t^i represents a distinct phase of biological development—such as gastrulation ($i = 1$), organogenesis ($i = 2$), or maturation ($i = 3$). The metric tensor g_{ij} encodes how dissimilar two developmental stages are, even if they occur at the same chronological age in different individuals. Unlike Euclidean distance, ds^2 quantifies *functional proximity*: two embryos are "close" if their gene-expression profiles, cell-fate maps, or morphogen gradients align at corresponding phases, irrespective of calendar time. Formally, $\Delta t^i = t^i - t'^i$ is the coordinate difference between stages of a current organism and a past one. The positive-definite matrix g_{ij} (analogous to spacetime curvature in general relativity) warps this space so that evolutionarily conserved transitions (e.g., neural tube closure) have small ds, while divergent paths (e.g., metamorphosis vs. direct development) yield large ds. This metric underpins all subsequent resonance: influence propagates most strongly between developmentally homologous states, enabling a species to "recognize" its own history across generations without relying on DNA alone. The construction ensures invariance under reparameterization of developmental time, making ds an intrinsic measure of biological similarity.

2. Morphic Field Accumulation

$$M^i(t^k) = \int K^i{}_j(t^k, t'^k)\, \phi^j(t'^k)\, d^3t'$$

(multiplication between $K^i{}_j(t^k, t'^k)$ and $\phi^j(t'^k)$; multiplication by differential measure d^3t'; dot product implied by Einstein summation over j)

The morphic field M^i at a given developmental coordinate t^k (a point in the 3D time space) is a vector field that aggregates contributions from the final adult forms $\phi^j(t')$ of *all past organisms*. Here, ϕ^j is the shape descriptor—a high-dimensional vector encoding organ positions, tissue boundaries, or Fourier coefficients of the body plan—at the endpoint of development for a historical individual. The integral sums over every past developmental trajectory t'^k, with each contribution weighted by the resonance kernel K^i_j (defined in Eq. 4). This is a **Fredholm integral equation of the second kind**, meaning M is defined in terms of an integral over itself via prior solutions—forming a self-consistent "memory" built from the ensemble of all ancestral outcomes. The kernel K linearly maps each ancestral adult form into the current embryo's coordinate system, so that, for example, a butterfly's wing pattern influences a developing pupa *only* through **homologous imaginal discs** (the shared precursor tissues). The measure d^3t' formally integrates over a continuum of past paths but, in practice, sums discretely over the finite population. Convergence requires K to decay rapidly (Gaussian, Eq. 3) to avoid divergence from infinite history. Thus, M^i emerges as a **collective, distributed memory**—not encoded in any single genome, but sustained across the species' temporal cloud, incrementally updated (Eq. 9) and continuously queried during ontogeny.

3. Gaussian Weight

$$G(ds^2) = \exp\left(-\frac{ds^2}{2\sigma^2}\right)$$

(no neighboring terms involving multiplication or dot product)

Here, **exp** denotes the exponential function e^x, where $e \approx 2.718$ is the base of natural logarithms. Thus, $G(ds^2) = e^{-\frac{ds^2}{2\sigma^2}}$ produces a value between 0 and 1, starting at $G(0) = 1$ and falling smoothly to near zero as ds^2 increases.

This exponential form generates a **bell curve** in the space of developmental distance ds^2. The curve is centered at $ds^2 = 0$ (perfect developmental alignment) and spreads symmetrically, with width controlled by σ. This shape embodies **fading memory**: influence from past organisms is strongest when their developmental stage exactly matches the current one ($ds^2 = 0$), and weakens gradually—never abruptly—as stages diverge. A small σ creates a narrow peak, so only nearly identical past forms contribute significantly; a large σ widens the bell, allowing broader, weaker resonance across varied developmental paths. The exponential ensures smooth, continuous decay, mirroring how real biological systems favor gradual transitions and robust convergence. Mathematically, this is the same function that governs heat spreading in all directions from a point source, with the total influence normalizing to $\int G(ds^2)\, d^3t' \approx (2\pi\sigma^2)^{3/2}$. In essence, G turns Sheldrake's idea of morphic resonance into a precise, graded attraction: past shapes pull hardest when they're developmentally "near," with far-off outliers ignored.

4. Resonance Kernel

$$K_{ij}(t^k, t'^k) = G(ds^2)\,\delta_{ij} + \kappa\,\partial_i\partial_j G(ds^2)$$

(multiplication between $G(ds^2)$ and δ_{ij}; multiplication between κ and $\partial_i\partial_j G(ds^2)$)

This kernel defines **how** past adult forms influence a developing embryo—combining **direct attraction** to historical endpoints with **smooth guidance** along well-traveled developmental paths.

The **first term,** $G(ds^2)\,\delta_{ij}$, pulls each phase of the current embryo directly toward the corresponding phase in past adults. Here, δ_{ij} is the **Kronecker delta**:

$$\delta_{ij} = \begin{cases} 1 & \text{if } i = j \\ 0 & \text{if } i \neq j \end{cases}$$

It ensures **phase-specific resonance**: gastrulation ($i = 1$) is influenced only by past gastrulation, organogenesis ($i = 2$) only by past organogenesis, etc. **Gastrulation** is the early embryonic stage when a hollow ball of cells (blastula) folds inward to form three germ layers—ectoderm (skin, brain), mesoderm (muscle, bone), endoderm (gut)—establishing the body's basic layout. No cross-talk between phases means a wing pattern never misguides leg development.

The **second term,** $\kappa\,\partial_i\partial_j G(ds^2)$, adds **path guidance.** Here, $\partial_i\partial_j G$ is the **second derivative** (Hessian) of the Gaussian G—measuring the **curvature** of the influence landscape. In physics, second derivatives govern **acceleration**; here, they sculpt **smooth channels** or **grooves** in developmental space, steering growth along trajectories reinforced by countless ancestors.

Although the adult shape ϕ^j (the high-dimensional vector encoding final organ positions and tissue boundaries) does not appear in this equation, it is the **content** being transmitted. Equation 4 defines the **mechanism**—the wiring—while Eq. 2 applies it:

$$M^{i} = \int K^{i}{}_{j}\, \phi^{j}\, d^{3}t'$$

→ K_{ij} tells *how* to connect; ϕ^{j} supplies *what* to send.

Term	Role	Physical Analogy
$G\,\delta_{ij}$	Direct pull to past adult form	Spring snapping to a target
$\kappa\,\partial_i\partial_j G$	Path guidance via curvature	Skiing in a groomed gully

Together, they form a **biological GPS**: embryos are drawn to ancestral endpoints while gliding along **historically smoothed highways**. This dual system explains **canalization**—robust convergence to the same adult form via slightly variable, resilient routes.

5. Geodesic in Morphic Cloud

$$\exp_p(v) = \gamma(1), \quad \frac{D\gamma}{d\lambda} = M^i(\gamma)\frac{\delta}{\delta t^i}$$

(multiplication between $M^i(\gamma)$ and $\frac{\delta}{\delta t^i}$; dot product implied by Einstein summation over i)

This pair of equations describes **how an embryo grows**: not by following a rigid script, but by traveling along the **smoothest, most-reinforced path** through developmental time—guided by the collective memory of all past members of its species.

Imagine developmental time as a **three-dimensional landscape** (t^1, t^2, t^3) where each point represents a specific combination of biological phases (e.g., gastrulation, organ budding, maturation). The embryo starts at the origin $(t^k = 0)$, the zygote. Its journey to adulthood traces a curve $\gamma(\lambda)$, where λ is a normalized "biological clock" running from 0 (fertilization) to 1 (reproductive maturity).

The **first equation**, $\exp_p(v) = \gamma(1)$, says:

> Start at point p (the zygote), take an initial direction v in this time-space, **and follow the natural, straight-line path defined by the morphic field.** After traveling a distance of 1 along this path, you arrive at $\gamma(1)$—the adult form.

This "straight line" is a **geodesic**—the shortest, least-effort route on a curved surface, like a great circle on a globe.

The **second equation** tells *how* to stay on the geodesic:

$$\frac{D\gamma}{d\lambda} = M^i(\gamma)\,\frac{\delta}{\delta t^i}$$

> At every moment, the **velocity** of development $\left(\frac{D\gamma}{d\lambda}\right)$ must point **exactly in the direction of the morphic field** M^i at the current position γ.

- M^i is the **arrow field** built from all past adults (via Eq. 2)—it points where history says "this way works best."
- $\frac{\delta}{\delta t^i}$ are the basis directions in time-space (like unit vectors along each phase axis).
- The equation says: **move parallel to the field**—no turning unless the field itself curves.

There are **no extra correction terms** (no Christoffel symbols), meaning the field M *defines* the geometry. Development is **"straight" in the space where past forms are densest**—like water flowing downhill along a riverbed carved by prior streams.

In practice:

- The embryo explores slightly via noise (mutations, environment).
- But on average, it **snaps to the central tendency** of ancestral paths—the **principal trajectory** of the species.

Thus, **ontogeny does not replay phylogeny in calendar time**, but as the **dominant eigenmode** of the morphic system: the single most probable route through the cloud of all past journeys.

6. Action Functional

$$S[\gamma] = \int \left[\frac{1}{2} g_{ij} \dot{\gamma}^i \dot{\gamma}^j - V_{\text{gene}}(\gamma) + \kappa \, \delta\phi \cdot M(\gamma) \right] d\tau$$

Plain English meaning:
This equation calculates a single *total score* for any possible path an embryo might take from zygote to adult. The path, denoted γ, is a curve through a three-dimensional "developmental time" space—where each axis (t^1, t^2, t^3) tracks progress in a major biological phase (e.g., gastrulation, organ formation, maturation). The terms $\dot{\gamma}^i$ and $\dot{\gamma}^j$ are the *speeds* along these three axes—how fast the embryo advances in each phase at any moment. The score S is the sum of three contributions, added up continuously from start to finish. Nature effectively *chooses the path with the lowest total score*, guiding real development toward the smoothest, most genetically faithful, and historically proven route.

Why integrate? Why sum the terms inside the parentheses over τ?
The integral means: **evaluate the score at every instant along the journey and add it all up**. Development isn't judged by a single snapshot—it unfolds over time. At each moment (τ), the embryo has a position $\gamma(\tau)$ and a velocity $\dot{\gamma}(\tau)$ in developmental space. The expression inside the brackets computes an *instantaneous cost* (or reward) at that moment. Integrating over τ from fertilization to maturity totals the cost across the entire life cycle—like calculating the full price of a road trip by adding fuel, tolls, and wear at every mile. Without integration, we'd only score one fleeting instant; with it, we judge the *entire story* of growth.

The three terms, clearly explained:

1. **Smoothness cost** ($\frac{1}{2}g_{ij}\dot{\gamma}^i\dot{\gamma}^j$):

   This measures how smoothly the embryo moves through developmental phases. The g_{ij} (from Eq. 1) defines biological "distance"—some phase transitions are inherently longer or riskier. High speeds ($\dot{\gamma}$) in hard directions cost more, like accelerating sharply on a bumpy road. This term penalizes erratic or rushed development, favoring steady, coordinated progress.

2. **Genetic target reward** ($-V_{\text{gene}}(\gamma)$):

   DNA defines ideal checkpoints (e.g., "heart forms here"). V_{gene} is a landscape with deep valleys at these targets. Being in a valley *lowers* the score (a reward); being on a peak raises it (a penalty). The negative sign turns attraction into a benefit—like gravity pulling a ball downhill to reduce its total energy.

3. **Historical alignment reward** ($\kappa\,\delta\phi \cdot M(\gamma)$):

   This is the *resonance bonus*. $\delta\phi$ is how much the embryo's current shape deviates from the ancestral average at this exact developmental stage. $M(\gamma)$ is the species' collective memory field—pointing toward what past adults actually achieved. The dot product measures alignment: perfect match gives maximum reward (lowest score); mismatch raises it.
   Why call this "potential energy"? Because it works exactly like gravitational or electrical potential: the embryo "falls" toward configurations where history is densest, just as a ball rolls into a valley. The morphic field M defines a *landscape of past success*—being near the historical norm lowers the score, pulling development toward proven outcomes, much like potential energy guides a pendulum to its lowest point.

In everyday terms:
Think of planning a career. You want:

- Steady progress (no burnout or gaps),
- Hitting key milestones (degrees, skills),
- Following paths others in your family succeeded in.

The equation scores every possible life plan moment by moment, then sums the total. The winning path feels effortless because it's smooth, meets inherited goals, *and* aligns with ancestral success—like potential energy naturally guiding water downhill.

7. Resonance Score

$$\mathcal{R} = \int \delta\phi_i(t^2, t^3)\, M^i(t^1, t^2, t^3)\, dt^2\, dt^3$$

(dot product via Einstein summation over i; integration over later phases t^2, t^3)

Plain English meaning:
$\mathcal{R}$ is a **compatibility score** that measures how well a *trial body plan* during development aligns with the **collective memory** of all past adults from the same species — *but only those that followed the same early path.*

Here's how it works:

- The embryo has already progressed to a fixed early stage (e.g., blastula, fixed at t^1).
- It now *tests* different possible shapes $\phi(t^2, t^3)$ for later phases (organogenesis, maturation).
- $\delta\phi_i = \phi_i^{\text{trial}} - \phi_i^{\text{reference}}$ is the **deviation vector** — how much the trial shape differs from a baseline (say, the current average).
- $M^i(t^1, t^2, t^3)$ is the **morphic field arrow** at that same point in developmental space — pointing toward the *average adult outcome* achieved by all ancestors who passed through the same early stage t^1.

The key interaction:

$$\delta\phi_i\, M^i \quad \text{(summation over } i\text{)}$$

This is a **dot product** — it measures **how parallel** the deviation $\delta\phi$ is to the historical direction M.

- If $\delta\phi$ points *with*M → high positive $\mathcal{R}$ → strong resonance → "this path worked before."
- If opposite → negative $\mathcal{R}$ → "avoid this — ancestors didn't go here."
- If perpendicular → $\mathcal{R} \approx 0$ → neutral.

The integral averages this alignment **across all later stages** (t^2, t^3), asking: *"Does this whole future trajectory feel familiar to the species?"*

Why no t^1 integration? Causality. Only past organisms that *matched the current embryo's early development* (t^1) are allowed to vote. Others are irrelevant.

Biological analogy:
Like a choir singing in harmony. Each cell "hears" the morphic field M (the sheet music passed down generations) and adjusts its note (ϕ) to match. The resonance score $\mathcal{R}$ is the **volume of harmony** — loudest when everyone sings the ancestral tune.

In practice, cells don't compute integrals — but molecular feedback loops **locally maximize** $\mathcal{R}$, like ants following a pheromone trail. Isolated embryos (weak M) wander; group-reared ones lock in fast.

Thus, $\mathcal{R}$ turns species memory into a **predictive compass**: "This way feels right because *we've been here before.*"

8. Equation of Motion

$$\frac{D^2\gamma^k}{d\tau^2} + \Gamma^k_{ij}\dot{\gamma}^i\dot{\gamma}^j = -\partial^k V_{\text{gene}} + \kappa K^k{}_l\,\delta\phi^l$$

This is **the law of biological motion**—how an embryo accelerates through developmental space to become an adult. Think of the embryo as a **sled sliding down a 3D hill** shaped by two forces: **genes** and **ancestral memory**. The path is $\gamma^k(\tau)$, where $k = 1, 2, 3$ labels early, middle, and late development, and τ is "growth progress" from 0 (egg) to 1 (adult).

Left Side: Inertia and the Shape of the Hill

- $\dfrac{D^2\gamma^k}{d\tau^2}$ → **Acceleration**

  How fast the embryo is changing direction in developmental space. Example: a limb bud suddenly lengthening = high acceleration in "limb growth" direction.

- $\Gamma^k_{ij}\dot{\gamma}^i\dot{\gamma}^j$ → **Curvature force**

  Γ comes from the metric g_{ij} (Eq. 1). It acts like **guardrails** on a mountain road. You can't form a brain before a skull—Γ creates a **pushback** if you try.

  $\dot{\gamma}^i$ is current speed through stage i. High speed through organ formation? Γ applies a **corrective nudge** to keep sequence logical.

Metaphor: The left side is the **sled's momentum and the hill's banking**. It resists flips and enforces realistic turns.

Right Side: The Two Forces

Force 1: Genetic Pull

$-\partial^k V_{\text{gene}}$

- V_{gene} is a **landscape of valleys** drawn by DNA.
  Deep valley at "four limbs," "one heart," "eyes on front."
- $-\partial^k V$ = **downhill slope** in direction k.
  The embryo is a ball **rolling toward the nearest genetic X-mark.**

Example: A chick embryo with a wing bud slightly early feels a **chemical tug** to slow down—Hox genes enforce timing.

Force 2: Resonance Force — Now with Full Kernel

$\kappa\, K^k{}_l\, \delta\phi^l$

This is the **ancestral push**. It uses the **full resonance kernel from Equation 4**:

$K_{ij}(t^k, t'^k) = G(ds^2)\,\delta_{ij} + \kappa\,\partial_i\partial_j G(ds^2)$

Let's **plug this in** and **expand** the resonance term:

$\kappa\, K^k{}_l\, \delta\phi^l = \kappa\left[G(ds^2)\,\delta^k_l + \kappa\,\partial_l\partial^k G(ds^2)\right]\delta\phi^l$

This splits into **two clear actions:**

Action 1: "Go to this shape"

$\kappa\, G(ds^2)\, \delta\phi^k$

- $G(ds^2) = \exp\left(-\frac{ds^2}{2\sigma^2}\right)$ → **Influence strength** (Eq. 3)
  How close this ancestor's path was to yours. Close = $G \approx 1$, far = $G \approx 0$.
- $\delta\phi^k$ → **Shape error in direction** k
  Example: "Your arm is 10% too short compared to adult average."
- **Result**: A **direct pull** toward the ancestral adult shape, **scaled by closeness**.

Biology: A human fetus with short arms feels **growth hormone surges** in arm buds—pushed to match the **average arm length** of 8 billion ancestors.

Metaphor: A **giant hand** made of all past adults, gently **stretching** your limbs to match theirs.

Action 2: "Take this exact route"

$\kappa^2\, (\partial_l \partial^k G)\, \delta\phi^l$

- $\partial_l \partial^k G$ → **Second derivative of the Gaussian**
  This measures **how the influence landscape curves** as you move in developmental directions l and k.
- **Result**: Creates **grooves or channels** in the hill—**paths worn smooth by past embryos**.
- **Biology**: Not just "have long arms," but **grow them at the same rate** as ancestors. If arms lag, this term **accelerates arm growth** while **slowing torso** to maintain proportion.

Metaphor: A **bobsled run** carved by billions of prior sleds. Your sled is **pushed into the groove**—not just to the finish, but **along the proven track**.

Putting the Resonance Force Together

$$\underbrace{\kappa G\delta\phi^k}_{\text{Pull to adult shape}} + \underbrace{\kappa^2(\partial_l\partial^k G)\delta\phi^l}_{\text{Push along ancestral path}}$$

- **First term**: "**Match the final photo** of adulthood."
- **Second term**: "**Follow the exact sequence** that got them there."

κ controls **how much history overrides genes**. In butterflies, high κ → rigid metamorphosis timing. In humans, lower κ → flexible brain growth.

Full Picture

Left side = sled inertia + hill banking
Right side =

- **Genes**: "Roll to the X"
- **Ancestors:** "Stretch to our size **and** slide in our groove"

At balance: **smooth, rapid, robust development**.

Damage? Groove reroutes.
Mutation? New groove forms if viable.
Convergence? All sleds end in the same adult basin.

This is **development as guided physics**—with memory.

9. Morphic Field Update

$$\frac{\partial M^i}{\partial t^1} = \kappa\,\delta^{(3)}\left(t^k - \gamma^k(\tau)\right)\phi^i(\gamma(\tau))$$

(multiplication between κ and $\delta^{(3)}$; multiplication between the delta and ϕ^i)

Plain English meaning:
This equation describes exactly how and when the species' collective memory is strengthened by each new generation. The morphic field M^i is updated only by organisms that successfully reach reproductive adulthood, and it is updated only along the precise developmental path they actually took.

Here is the step-by-step picture:

1. A new organism develops from zygote to reproductive adult, tracing its own curve $\gamma(\tau)$ through the 3D developmental space (where τ runs from 0 = fertilization to 1 = reproductive maturity).
2. Only if it completes the whole journey successfully does it get to contribute. Its final adult body plan (organ positions, tissue boundaries, etc.) is recorded as the vector ϕ^i.
3. That finished blueprint is then deposited back into the morphic field exactly along the path it followed — nowhere else.

   The term in parentheses,

   $$\left(t^k - \gamma^k(\tau)\right)$$

   is the distance (in all three phase directions) between

   - a point t^k we are examining in the big 3D space, and
   - the point $\gamma^k(\tau)$ where this particular embryo actually was at clock time τ.

   The 3D Dirac delta $\delta^{(3)}\left(t^k - \gamma^k(\tau)\right)$ is an infinitely narrow spike that is non-zero only when that distance is exactly zero in all three dimensions at once.

In other words:
"Add this adult's final shape to the memory only at the exact coordinates this individual passed through, and do it with laser precision — no smearing, no influence on paths it never took."

4. The left side $\frac{\partial M^i}{\partial t^1}$ tells us the reinforcement is applied gradually and primarily during the earliest phase (t^1), because that is when embryos are most sensitive to guidance.
5. The constant κ is a volume knob: it controls how much influence each new adult exerts. In real populations it is kept naturally small by population size and generation overlap.
6. Only proven lives count. Embryos that die or fail to reproduce drop nothing into the field — their paths fade away.

Clear analogy
Think of the 3D developmental space as a huge dark forest.
Every successful organism walks a glowing trail from the entrance (zygote) to the far exit (reproductive adult).
When it finally steps out the exit, it turns around and paints its finished adult portrait along every inch of the exact trail it just walked — using an infinitely thin brush (the Dirac delta).
Future embryos walking through the forest now see slightly brighter paint wherever many successful ancestors have walked before, and they are gently pulled toward those brighter, proven routes.

In discrete (real) populations the equation simply becomes an update rule:

$$M^i(\text{point on path}) \leftarrow M^i + \kappa \times (\text{final adult shape of this individual})$$

Big picture
No DNA change is required for the species to "learn." Every completed, successful life sharpens the shared memory, making the next generation a little more likely to follow the best routes. This is the mathematical mechanism behind the rapid, non-genetic adaptations sometimes seen in isolated colonies or laboratory cultures: the field remembers, and it teaches.

Mathematical Annex: The Mathematics of Speciation in Morphic Resonance

(Equations 1–6: How new species are born)

1. Daughter Field Initialization

$$M_D^i(t_{\text{split}}^k) = M_P^i(t_{\text{split}}^k) + \epsilon\, \Delta M_{\text{mutation}}^i(t_{\text{split}}^k)$$

This equation marks the **birth moment of a new species**—the instant when a daughter lineage breaks away from its parent, carrying a nearly identical memory but with a tiny twist that sets evolution in motion. Imagine the parent species as a vast library of developmental blueprints, accumulated over millions of generations. At the splitting point t_{split}^k, the daughter starts with an exact copy of the parent's morphic field M_P^i at that precise developmental coordinate $t^k = (t^1, t^2, t^3)$—early embryo, organ formation, maturation.

The field M^i is a vector arrow at every point in developmental space, pointing toward the average successful adult form. The daughter inherits this arrow fully: same heart position, limb proportions, neural wiring biases. But evolution demands difference. The term $\epsilon\, \Delta M_{\text{mutation}}^i$ is a **small mutational nudge**—a perturbation in the memory field itself.

Here, ϵ is a tiny scalar (e.g., 0.01 or less), ensuring the change is subtle, not catastrophic. $\Delta M_{\text{mutation}}^i$ could arise from:

- A genetic mutation altering early gene expression, shifting the field's starting point.
- Environmental stress (e.g., isolation on an island) that favors rare variants surviving to adulthood.
- Stochastic noise in a small founder population, amplifying a deviation.

Think of it like photocopying a book but with a slight ink smudge on the first page. The daughter library opens with 99.9% identical pages, but that smudge propagates. In biological terms, this is the **founder event** in allopatric speciation: a few finches blow to the Galápagos. Their collective M_D starts as the mainland M_P, plus a nudge toward longer beaks from drought survivors.

The indices i align homologous structures: a nudged wing field in birds doesn't affect unrelated gut fields. The coordinate t^k_{split} is critical—speciation often hinges on early divergence (e.g., regulatory genes like Hox). If the nudge is at t^1 (embryo), it cascades; at t^3 (maturation), it's superficial.

Why add to the field, not DNA? Morphic resonance posits memory is non-local. The nudge ΔM could be a viable adult form from a mutant that "resonates" slightly off the main cluster, pulling the daughter average. In practice, the first daughter generation develops under the old M_P, but survivors update M_D, ratcheting the change.

This initialization prevents instant chaos. Without it, daughters would collapse back to parents via strong resonance. With it, a **seed of novelty** is planted. In fossil terms, this is the **anagenesis-cladogenesis boundary**: gradual shift becomes branching when ϵ accumulates.

Mathematically, this is a **bifurcation point**. For $\epsilon = 0$, $M_D = M_P$ forever—stasis. For $\epsilon > 0$, paths diverge (see Eq. 3). In simulations, small ϵ yields microevolution; larger (e.g., polyploidy) yields macro.

In short: speciation isn't a genetic big bang. It's a **gentle fork in the memory road**, starting with a whisper of difference in the collective past.

---

2. Divergence Kernel

$$K^{i}_{D\leftarrow Pj} = G_{\text{div}}(ds^2_{PD})\,\delta^i_j\,e^{-\lambda(t^1 - t_{\text{split}})}$$

This equation governs the **fading echo of the parent** in the daughter's development—a lingering influence that weakens over time and as forms drift apart. After initialization (Eq. 1), the daughter is not isolated. Early on, the parent's ghosts still whisper guidance. But speciation requires independence. The divergence kernel $K_{D\leftarrow P}$ serves as the **communication link** from the parent field to the daughter embryo.

The core is $G_{\text{div}}(ds^2_{PD})$, a **Gaussian function**—explicitly an exponential of the form $\exp\left(-\frac{ds^2_{PD}}{2\sigma^2_{\text{div}}}\right)$—that weights **inter-species distance**. The quantity ds^2_{PD} measures biological mismatch between parent and daughter developmental paths. It uses the metric g_{ij}, but now compares entire trajectories after the split. When ds_{PD} is small, daughter populations still follow the same developmental "acts." When it is large, body plans diverge significantly—for example, fins evolve into legs.

The term δ^i_j ensures **homology mapping**: resonance from a parent's wing affects only the daughter's homologous wing structure, if one exists. The exponential $e^{-\lambda(t^1-t_{\text{split}})}$ controls **time decay**. Here, t^1 represents calendar time since the split (not developmental time), and λ is the decay rate.

- **Fast** λ (large numerical value, e.g., $\lambda = 1.0$): The parent influence drops sharply. For example, it falls to $e^{-1} \approx 0.37$ in one generation. This leads to quick independence. A clear example is the rapid evolution of **hybrid sterility**, where the daughter lineage becomes reproductively isolated almost immediately.
- **Slow** λ (small numerical value, e.g., $\lambda = 0.01$): The parent influence lingers across many generations. For example, it remains at $e^{-0.01} \approx 0.99$ after one generation. This allows prolonged overlap. A classic example is the **ring species** of Ensatina salamanders in California. These populations form a geographic ring around the Central Valley. Neighboring groups interbreed freely due to strong resonance and slow decay. However, where the ring closes in Southern California, the two endpoint populations—*E. eschscholtzii* (blotched) and *E. klauberi* (plain)—meet but do not interbreed, despite being connected by a continuous chain of interbreeding populations. Here, λ is small: the parent echo fades only over **tens of kilometers and thousands of generations**, enabling gradual divergence without sudden isolation.

Imagine the parent as a radio station broadcasting adult forms. At the moment of split, the daughter tunes in fully ($G \approx 1$, exp ≈ 1). As generations pass and t^1 increases, the signal fades exponentially—like driving away from the tower. At the same time, as daughter forms evolve and ds_{PD} grows, the Gaussian exponential muffles irrelevant advice.

Biologically, this explains **introgression**. Early-generation hybrids are viable because the parent kernel remains strong. Later, hybrids become incompatible as the influence weakens. In Darwin's finches, ground and tree finches share early beak resonance when ds is small. But seed-crushing versus probing adaptations cause the influence to decay over time.

The kernel is **directional** ($D \leftarrow P$): the daughter feels the parent, but the parent remains unchanged. The indices ij allow **part-specific decay**: core body plan elements like the spine decay more slowly than peripheral traits like the beak.

Both exponentials—Gaussian for form distance and pure exponential for time—ensure a smooth, rapid drop-off without abrupt switches.

In practice, the divergence kernel operates by integrating contributions from all past parent individuals. For a daughter embryo at a specific developmental coordinate t^k, the kernel sums the adult forms ϕ_P of every ancestor in the parent lineage. Each contribution is weighted by similarity in developmental path and by time since the split. Early after speciation, when the daughter population is small and its own morphic field is weak, the parent's resonance dominates guidance. As generations pass and the daughter's own successful adults accumulate, their collective field strengthens and gradually supplants the fading parental influence (see Eq. 3).

This kernel acts as the **bridge during parapatric speciation**—a mode of speciation in which daughter populations diverge while remaining in **partial geographic contact** (e.g., along an environmental gradient or at the edges of a range), allowing limited gene flow that gradually diminishes as morphic resonance fades. Without it, daughter lineages would be vulnerable. With it, they achieve gradual emancipation. The threshold for recognizing a "true" species occurs when the kernel approaches zero (links to Eq. 4).

In short: the parent's voice does not vanish overnight. It **fades like an echo in a canyon**, giving the daughter time to find its own song.

3. Daughter Field Evolution

$$\frac{\partial M_D^i}{\partial t^1} = \kappa_D K_{Dj}^i \phi_D^j + \kappa_{D \leftarrow P} K_{D \leftarrow Pj}^i \phi_P^j + \eta \xi^i(t^k)$$

This equation governs **how the daughter species' morphic field evolves over generations** after the initial split. The morphic field $M_D^i(t^k)$ is a vector-valued function that, at every point $t^k = (t^1, t^2, t^3)$ in developmental space (early embryo, organ formation, maturation), stores the **average adult body plan** achieved by past individuals who passed through that point. The index i labels components of the body plan (e.g., spine curvature, limb length).

The left side, $\frac{\partial M_D^i}{\partial t^1}$, is the **time derivative** of the daughter field with respect to **calendar time**t^1 (measured in generations). It tells us how the collective memory is updated as new adults are born.

The right side contains **three additive contributions**:

1. Own-kind reinforcement

$$\kappa_D K_{Dj}^i \phi_D^j$$

- ϕ_D^j: The **final adult form** (a vector) of a specific daughter individual that survived to reproduce.
- K_{Dj}^i: The **daughter resonance kernel**, a matrix that measures **developmental similarity** between the current embryo's path and the path taken by past daughter adults. It is constructed from a Gaussian $G_D(ds^2)$ centered on the daughter's own history.

- κ_D: A **positive scalar learning rate**. In small populations, $\kappa_D \sim 1/N_D$ (where N_D is population size), so each survivor has greater influence; in large populations, the effect of any single individual is diluted.

Interpretation: This term **accumulates successful daughter forms**, weighted by how closely their developmental trajectories match the current one. It is the mechanism by which the daughter field **self-reinforces** its emerging identity.

2. Parent echo

$\kappa_{D\leftarrow P} K^i_{D\leftarrow Pj} \phi^j_P$

- ϕ^j_P: The **final adult form** of a parent-lineage individual.
- $K^i_{D\leftarrow Pj}$: The **divergence kernel** defined in Eq. 2:

  $K^i_{D\leftarrow Pj} = G_{\text{div}}(ds^2_{PD})\, \delta^i_j\, e^{-\lambda(t^1 - t_{\text{split}})}$

  Here $G_{\text{div}}(ds^2_{PD}) = \exp\left(-\frac{ds^2_{PD}}{2\sigma^2_{\text{div}}}\right)$ is a **Gaussian function** that **decreases smoothly as the developmental mismatch ds^2_{PD} between parent and daughter trajectories grows larger**. The term ds^2_{PD} is the **squared biological distance** (computed with the metric g_{ij}) between the full developmental path of a parent individual and that of the current daughter embryo. Thus, the Gaussian ensures that only parent forms whose entire growth sequence remains biologically close contribute meaningfully; distant forms are effectively ignored. This Gaussian decay is then multiplied by the **exponential time decay**$e^{-\lambda(t^1 - t_{\text{split}})}$ and the homology mapping δ^i_j.

- $\kappa_{D\leftarrow P}$: A **scalar** that starts near κ_D at $t^1 = t_{\text{split}}$ and decreases (often modeled as $\kappa_{D\leftarrow P} = \kappa_D e^{-\mu(t^1 - t_{\text{split}})}$).

Interpretation: This term represents the **fading influence of the parent species**. Early after speciation, it stabilizes development and enables viable hybrids. As time passes and forms diverge, its contribution vanishes.

3. Innovation noise

$\eta\xi^i(t^k)$

- $\xi^i(t^k)$: A **random fluctuation** at every point (t^k) in developmental space. Mathematically, it is **white Gaussian noise**:
  - **Gaussian**: each value follows a normal (bell-shaped) distribution with mean 0 and fixed variance.
  - **White**: the noise is **uncorrelated in time and space**—the value at one developmental point has **no statistical relationship** with the value at any other point. Formally:

    $\langle \xi^i(t^k) \rangle = 0, \quad \langle \xi^i(t^k)\xi^j(t'^k) \rangle = \delta^i_j \delta^3(t^k - t'^k)$

    The **Kronecker delta**δ^i_j means independence across body axes; the **Dirac delta**$\delta^3(t^k - t'^k)$ means independence across developmental stages.
- η: A **positive scalar** that sets the **strength** of the noise.
  - **High** η: large random kicks → common during population crashes, extreme environments, or genetic bottlenecks.
  - **Low** η: tiny wiggles → typical in large, stable populations.

Interpretation:
This term **injects small, random changes** into the morphic field at every generation. Most of these changes are harmless or harmful and disappear quickly. But **rarely**, a random deviation produces a **beneficial trait**—say, a slightly longer beak that reaches deeper seeds. If that individual survives and reproduces, its adult form ϕ^j_D is added via **Term 1** in the next generation. Over time, **useful noise becomes structure**.

Think of it like **mutation in DNA**, but at the level of **developmental memory**. It is the **spark of novelty** that prevents the daughter lineage from being a perfect copy of the parent.

In plain terms:
Evolution needs **room to experiment**. This term is that room—a quiet background of random variation that occasionally hits on something better.

Mathematical Structure

The equation is a **stochastic integro-differential equation** (the kernels K implicitly involve integrals over past trajectories). In discrete form, each generation adds:

$$M_D^i(t^1 + \Delta t) = M_D^i(t^1) + \Delta t\,[\text{Term 1} + \text{Term 2} + \text{Term 3}]$$

where $\Delta t = 1$ generation.

Biological Dynamics

Phase	Dominant Term	Outcome
Immediate post-split	Parent echo	Hybrid viability, developmental stability
Intermediate	Own-kind + noise	Rapid adaptation, canalization of new traits
Late	Own-kind only	Full reproductive isolation, stable daughter attractor

The equation explains **punctuated equilibria**, **adaptive radiations**, and **hybrid zone dynamics**. When noise generates a viable form outside the parent resonance basin, and selection amplifies it via Term 1, the daughter field **locks in** a new stable state—speciation complete.

In short: the daughter field is not static inheritance. It is a **dynamic, self-organizing memory** shaped by legacy (parent), consensus (own-kind), and chance (noise).

4. Speciation Threshold

$$\mathcal{O}_{DP}(t^1) = \frac{\int M_D^i M_P^i \, d^3t}{\sqrt{\int M_D^j M_D^j \, d^3t \cdot \int M_P^k M_P^k \, d^3t}} < \theta_c$$

This equation identifies the **tipping point** at which a daughter lineage becomes a truly separate species—one that can no longer merge back with its parent, even if they come into contact again.

Think of the morphic field M as a **map of developmental instructions**—a set of arrows, one at every stage of growth, pointing toward the average adult form that has worked in the past. Both parent and daughter have such maps. The symbol $\mathcal{O}_{DP}$ measures **how aligned these two maps still are** after the split.

How the overlap is calculated

The formula is a **normalized inner product** over the entire developmental space, analogous to the cosine similarity between two vectors—but here applied continuously across all growth stages.

- **Numerator: the inner product between fields**

  $$\int M_D^i M_P^i \, d^3t$$

  - At each point in developmental space (embryo → organ formation → maturation), compute the **component-wise simple multiplication** $M_D^i \cdot M_P^i$ (i.e., multiply corresponding arrow components: spine alignment × spine alignment, limb length × limb length, etc.).
  - Sum (integrate) these products over the full 3D developmental volume d^3t.

  - This is **not** a dot product at a single point—it is the **integral of local scalar products**, yielding a single number that grows larger when arrows consistently point in the same direction across development.

- **Denominator: normalization via field magnitudes**

$$\sqrt{\int M_D^j M_D^j \, d^3t \cdot \int M_P^k M_P^k \, d^3t}$$

  - For each field, compute the **squared magnitude** at every point via simple multiplication of a field with itself: $M_D^j M_D^j$ and $M_P^k M_P^k$ (i.e., sum of squared components of each arrow).
  - Integrate each over developmental space to get the total "energy" or "density" of past forms in that lineage.
  - Take the square root of their product to normalize.
  - This ensures $\mathcal{O}_{DP}$ is scale-invariant: a larger population (denser field) doesn't artificially inflate similarity.

- **Result**: $\mathcal{O}_{DP}$ is the **cosine of the angle** between the two fields in function space—1 if perfectly aligned, 0 if orthogonal.

The critical threshold

Speciation is declared when $\mathcal{O}_{DP} < \theta_c$ (e.g., 0.7). Below this, shared resonance is too weak to guide hybrid development successfully—**reproductive isolation** emerges naturally.

Visual analogy

Two glowing trails in developmental space. Early on, they overlap heavily—hybrids follow either path. As divergence proceeds (via Eqs. 2–3), trails separate. When **inner product falls** and **cosine similarity drops below** θ_c, the trails are distinct: no cross-guidance remains.

Real-world examples

- **Cichlids**: Color morphs diverge fast, but core body plans (high inner product) allow gene flow until structural shifts reduce $\mathcal{O}_{DP}$.
- **Crow hybrid zone**: Narrow zones = low $\mathcal{O}_{DP}$, few viable hybrids; wide zones = higher inner product, blending.

Why this matters

This is a **rigorous, quantitative species boundary**:

- Uses **simple multiplication** locally (per point, per trait).
- Uses **inner product integration** globally to measure alignment.
- Normalizes via **self-inner products** to remove population bias.

It marks the moment the **ancestral template loses influence**—not by decree, but by measurable silence in the shared developmental signal.

5. Form Bifurcation

$$\phi^i_D = \sqrt{1-p}\,\phi^i_P + \sqrt{p}\,\phi^i_{\text{novel}} + \sqrt{p(1-p)}\,\Phi^{ij}_{\text{entangle}}\phi^j_P$$

This equation describes **how a new trait appears in the daughter species—not as an isolated change, but as a coordinated shift that remains biologically viable because it is intertwined with the ancestral form.**

Imagine the adult body plan ϕ as a **vector of measurements**—limb length, beak depth, muscle thickness, etc.—one component (i or j) for each trait. The daughter's final form ϕ^i_D is built from three weighted contributions that blend together like ingredients in a recipe.

The three ingredients

1. **Ancestral core**

   $\sqrt{1-p}\,\phi^i_P$

   - This is the **parent's trait value** (ϕ^i_P), scaled by $\sqrt{1-p}$.
   - When $p = 0$, this term is 1—full inheritance, no change.
   - As p grows, the ancestral influence shrinks smoothly.

2. **Pure novelty**

   $\sqrt{p}\,\phi^i_{\text{novel}}$

   - A **brand-new trait value** (e.g., a longer beak tip), scaled by $\sqrt{p}$.
   - When $p = 1$, this dominates—radical innovation.
   - p acts like a dial: small values = minor tweak; large = major leap.

3. **Entanglement bridge**

   $\sqrt{p(1-p)}\,\Phi^{ij}_{\text{entangle}}\phi^j_P$

   - This is the **coordination term**—the reason a longer beak doesn't appear alone but comes with stronger jaw muscles.
   - Φ^{ij} is a **mixing matrix**: off-diagonal entries link related traits (e.g., beak depth → jaw strength).
   - The weight $\sqrt{p(1-p)}$ peaks at $p = 0.5$—maximum blending during transition.
   - It ensures novelty **resonates in context**, not as a random bolt-on.

Why the square roots and cross term?

The weights are chosen so that **total contribution sums to 1 in a probabilistic sense**, and the cross term introduces **constructive interference**—much like light waves combining in phase. This mimics **quantum superposition**: the new form is a coherent blend, not a patchwork. Without entanglement, a big beak on a weak jaw would fail (a "hopeful monster"). With it, changes are **holistic and functional**.

Real-world example: Darwin's finches

- **Parent**: medium beak, moderate crushing force.
- **Novelty**: deeper beak for cracking hard seeds (ϕ_{novel}).
- **Entanglement**: deeper beak pulls stronger muscles via Φ^{ij}.
- Result: a viable seed-crushing specialist, not a bird with a beak it can't support.

In plain terms

Evolution doesn't swap parts like Lego. It **remixes the old with the new in correlated ways**. The entanglement term is the glue that keeps innovation **biologically realistic**—new forms **dance with ancestral ghosts**, not stumble alone.

In short: viable novelty isn't random—it's **orchestrated coherence** between past and future.

6. Morphic Cloud Partition

$$\frac{D^2\gamma_D^k}{d\lambda^2} = \kappa_D\, K_{Dl}^k\, \delta\phi_D^l - \kappa_{D\leftarrow P}\, \nabla^k \mathcal{O}_{DP}$$

This equation describes how an embryo's developmental path actively diverges from its parental lineage, guided by two complementary forces: one that pulls it toward the proven forms of its own emerging species, and another that pushes it away from the parent's trajectory once sufficient difference has arisen.

Picture the embryo as a bead sliding along a flexible wire through a three-dimensional landscape of possible body plans. This wire is the developmental trajectory, denoted $\gamma_D^k(\lambda)$:

- k indexes the growth stage or body feature (e.g., limb bud, beak depth, neural tube),
- λ runs from 0 (embryo) to 1 (adult).

The left side, $\dfrac{D^2\gamma_D^k}{d\lambda^2}$, represents the curvature of this path — how sharply it bends at any given moment. A positive value steers it one way; a negative value steers it in the opposite direction.

The right side determines the direction of that bend.

The Two Forces — Simply Added

The right-hand side consists of two vector terms added together — no inner product between them, just direct vector summation:

$$\kappa_D\, K_{Dl}^k\, \delta\phi_D^l \quad + \quad -\kappa_{D\leftarrow P}\, \nabla^k \mathcal{O}_{DP}$$

Let us examine each in turn.

1. Pull Toward the Daughter Lineage

$\kappa_D \, K^k_{Dl} \, \delta\phi^l_D$

This term exerts an attractive force, guiding the embryo to follow paths that have historically led to reproductive success within its own lineage.

- $\delta\phi^l_D$ is a vector representing the deviation between the embryo's current position and the average adult form achieved by past successful individuals in the daughter lineage. It is the error signal: "You are this far off course."
- K^k_{Dl} is a matrix that evaluates relevance: "At this precise developmental stage k, which past traits l offer the most useful guidance?" It weights the influence of prior adults according to the similarity of their growth trajectories to the one currently unfolding.
- The expression $K^k_{Dl} \, \delta\phi^l_D$ involves summation over the repeated index l — the only place in the equation where contraction (an inner product) occurs. This is standard matrix-vector multiplication: the deviation is filtered through the relevance matrix to produce a directional correction.
- Finally, this result is scaled by κ_D, a positive scalar that governs the strength of the pull — stronger in small populations, where each survivor's influence is amplified, and weaker in large, stable ones.

In essence:
The embryo is drawn toward the well-trodden path carved by its own lineage's past winners, with guidance weighted by developmental similarity and amplified by population size.

It is as if the embryo hears a chorus of ancestral voices saying, "This way has worked before — follow it."

2. Push Away from the Parent Lineage

$$-\kappa_{D \leftarrow P} \nabla^k \mathcal{O}_{DP}$$

This term introduces active repulsion, ensuring that once the daughter lineage has begun to diverge, it does not drift back toward the parent's form — instead, differences are exaggerated.

Let's unpack this carefully.

Recall $\mathcal{O}_{DP}$

From Equation 4, $\mathcal{O}_{DP}$ is the overlap score — a scalar between 0 and 1 that quantifies how much developmental resonance still exists between daughter and parent morphic fields:

$$\mathcal{O}_{DP} = \frac{\text{integrated agreement between } M_D \text{ and } M_P}{\sqrt{\text{strength of } M_D} \cdot \sqrt{\text{strength of } M_P}}$$

- At the moment of splitting: $\mathcal{O}_{DP} \approx 1.0$ (nearly identical memories)
- At the speciation threshold: $\mathcal{O}_{DP} < \theta_c$ (e.g., 0.7)
- When fully isolated: $\mathcal{O}_{DP} \approx 0.0$

The Gradient $\nabla^k \mathcal{O}_{DP}$

This is a vector that answers a precise question:
"If I adjust the embryo's growth at this exact stage k, in which direction would $\mathcal{O}_{DP}$ increase the most?"

It points toward configurations that would make the daughter more parent-like. For example, if the parent lineage has evolved a medium-depth beak and the daughter is developing a deeper one, the gradient points toward shallower beak growth — because that change would increase overlap.

The Minus Sign

$-\nabla^k \mathcal{O}_{DP}$

This reverses the direction of the gradient. Now the vector points away from parent-like development.

The Strength Factor

$\kappa_{D \leftarrow P} \times (-\nabla^k \mathcal{O}_{DP})$

- $\kappa_{D \leftarrow P}$ is a scalar controlling the magnitude of repulsion.
- It begins at zero immediately after the split (no push needed — hybrids are viable).
- It increases only when $\mathcal{O}_{DP}$ falls below the critical threshold θ_c.
- A typical model is: $\kappa_{D \leftarrow P} = \kappa_0 \cdot \max(0, \theta_c - \mathcal{O}_{DP})$.

No summation occurs here — only scalar multiplication by a vector.

In essence:
The system identifies the developmental tweak that would most increase similarity to the parent, then actively steers the embryo in the opposite direction — but only after the lineages have diverged enough that shared resonance is fading.

This is character displacement made dynamic and precise.

When Does Repulsion Activate?

- Immediately after splitting: $\mathcal{O}_{DP} \approx 1.0$ → repulsion = 0 → hybrid offspring develop normally
- Intermediate phase: $\mathcal{O}_{DP} \to 0.8$ → weak repulsion begins to emerge
- Below threshold: $\mathcal{O}_{DP} < 0.7$ → strong, active repulsion → developmental paths rapidly diverge

Repulsion is not gradual drift. It is a threshold-triggered amplifier of difference.

A Concrete Example: Darwin's Finches in Sympatry

Two populations begin with identical medium-depth beaks. A small founder group, isolated briefly, returns with a slightly deeper beak due to a favorable mutation.

- $\mathcal{O}_{DP}$ drops from 1.0 to 0.85
- The gradient $\nabla^k \mathcal{O}_{DP}$ points toward medium-depth growth (the parent's form)
- Repulsion reverses this → pushes toward even deeper beaks
- Individuals with deeper beaks now crush hard seeds efficiently → survive and reproduce
- The original population retains the medium beak for softer seeds
- Food resources are partitioned → ecological separation reinforces reproductive isolation